Connections

DAVID WAUGH AND TONY BUSHELL

OXFORD
UNIVERSITY PRESS

OXFORD
UNIVERSITY PRESS

Great Clarendon Street, Oxford, OX2 6DP, United Kingdom

Oxford University Press is a department of the University of Oxford. It furthers the University's objective of excellence in research, scholarship, and education by publishing worldwide. Oxford is a registered trade mark of Oxford University Press in the UK and in certain other countries.

Key Geography Connections published in 1992
Key Geography Connections New Edition in 1997 by Stanley Thornes (Publishers) Ltd.
Key Geography New Connections (Third edition) published in 2001 and Key Geography New Connections (Fourth edition) in 2006 by Nelson Thornes Ltd.
Fifth edition published in 2014

British Library Cataloguing in Publication Data
Data available

978-1-4085-2317-9

1 3 5 7 9 10 8 6 4 2

Printed in Spain

Acknowledgements

Page make-up: GreenGate Publishing Services, Tonbridge, Kent
Illustrations: Kathy Baxendale, Nick Hawken, Angela Knowles, Gordon Lawson, GreenGate Publishing Services, Richard Morris, David Russell illustration, Tim Smith, John Yorke.

The publishers would like to thank the following for permissions to use their photographs:

Cover: hadynyah/iStockphoto; JordiRamisa/iStockphoto; NNehring/iStockphoto; **p4A:** James Marshall/Corbis; **p5B:** Owaki - Kulla/Corbis; **p5C:** Dave Martin/AP/Press Association Images; **p6A:** David Paterson Photography; **p6B, p6C:** Tony Waltham; **p6 (bottom left):** Tony Bushell; **p6 (bottom right):** Tony Waltham; **p7 (top left):** Tony Bushell; **p7 (top right):** ICCE Photolibrary/Mike Hoggett; **p8 (top left):** Fritz Polking/FLPA; **p8 (top right):** Michael Jenner/ Robert Harding; **p8 (bottom left):** H Richard Johnston/Getty; **p8 (bottom right):** beatrice prève/Fotolia; **p10B:** Tony Bushell; **p13D:** Robert Estall/Robert Harding; **p14B, p14C:** Tony Bushell; **p15E:** Frans Lanting, Mint Images/Science Photo Library; **p16A:** Jean Brooks/Robert Harding/Rex Features; **p17C:** English Heritage (Aerofilms Collection); **p18A:** John Giles/PA Archive/Press Association Images; **p20A:** Tony Bushell; **p22A:** Patrick Poendl/Shutterstock; **p22B:** Julian Cartwright/ Alamy; **p23D:** totophotos/Fotolia; **p24B:** The Photolibrary Wales/ Alamy; **p25C (top left):** Adam Burton/Alamy; **p25C (top right):** Francesco Carucci/Shutterstock; **p25C (bottom left):** Last Refuge/ Robert Harding; **p25C (bottom right):** Simon Warner; **p26A:** Panos Pictures/Jean Leo Dugast; **p27B:** Douglas Kirkland/Corbis; **p27C:** Douglas Peebles/Corbis; **p28A:** Impact Photos/Roger Scrutton; **p28B:** S. Forster/Alamy; **p28C:** Ken Walsh/Alamy; **p29E:** Alexander Raths/ Shutterstock; **p31B:** Richard Drury/Art Directors; **p31C:** Christopher Elwell/Shutterstock; **p31D:** Tony Waltham; **p31E:** Ann Taylor-Hughes/ Getty Images; **p32A:** Bob Gibbons/Eye Ubiquitous; **p33C:** John Farmar/ Art Directors; **p35B:** Royal Commission on the Ancient and Historical Monuments of Wales; **p36B:** Cultura Creative (RF)/Alamy; **p39D:** Eric James/Alamy; **p39E:** LOOK/Photoshot; **p40C:** Ocean/Corbis; **p41D (top left):** World Pictures/Photoshot; **p41D (top right):** Manoj Shah/Getty Images; **p41D (bottom left):** Alexander Tamargo/Getty Images; **p41D (bottom right):** StockShot/Alamy; **p42A**: Fujitsu Telecommunications Europe Ltd; Canon (UK) Ltd; PSL Images/Alamy (HP); imagebroker/ Alamy (IBM); Alistair Laming/Alamy (Apple); AFP/Getty Images (GSK); **p42B:** Darrin Jenkins/Alamy; **p44A:** Urbanmyt/Alamy; **p44B:** Cambridge Science Park; **p47D:** Mark James/Porthcawl Harbourside CIC; **p48A:** Christopher Herwig/Getty Images; **p49B:** Photoshot/Look; **p49C:** Urbanmyth/Alamy; **p49D:** Brenda Prince/Photofusion; **p51D:** Manfred Gottschalk/Getty Images; **p51E:** Ross Graham Photography; **p52A:** David Paterson Photography; **p52B:** Dr Morley Read/Science Photo Library; **p52C:** Simon Warner; **p52D:** Christiana Carvalho/ FLPA; **p52E**: Tony Bushell; **p52F:** Panos Pictures/Sean Sprague; **p56A:** artyzan12/Fotolia; **p57F:** David Bathgate/Corbis; **p58B:** Skyscan/Corbis; **p58C (top left):** D. Hurst/Alamy; **p58C (top right):** shotsstudio/ Fotolia; **p58C (middle left):** Nadino/Shutterstock; **p58C (middle right):** mamahoohooba/Shutterstock; **p58C (bottom left):** Ryan McVay/Photodisc 75 (NT); **p58C (bottom right):** Sally Greenhill; **p61C:** AFP/Getty Images; **p65C:** Photofusion/Molly Cooper; **p68A:** Pal Teravagimov/Shutterstock; **p69B:** Photoshot/Tibor Bognar; **p69C:** Colette3/Shutterstock; **p69D:** Jörg Hackemann/Fotolia; **p70A:** Robert Harding Picture Library Ltd/Alamy; **p70B:** Worldwide Picture Library/ Alamy; **p70C:** Michele Burgess/Alamy; **p72A:** Alain Le Garsmeur/Alamy; **p72B:** Cuboimages/Photoshot; **p72C:** JLImages/Alamy; **p74B: Vito** Arcomano/Alamy; **p74C:** Chris Rowley; **p75D:** Neil Emmerson/Robert Harding; **p75E:** David Pearson/Alamy; **p75F:** LOOK Die Bildagentur der Fotografen GmbH/Alamy; **p75G:** Imagebroker/Alamy; **p76A:** Tao/Robert Harding; **p76B:** Peter MacDiarmid/Rex Features; **p77D:** Gamma-Rapho via Getty Images; **p78B:** Philip Game/Alamy; **p80B:** Still Pictures/Robert Harding; **p81C:** Dinodia Photos/Alamy; **p81D:** Haytham Pictures/ Alamy; **p82A, p82B:** Chris Rowley; **p82C:** Gavin Hellier/Alamy; **p82D, p83E, p83F:** Chris Rowley; **p85C:** Dinodia Photos/Alamy; **p88A:** Skyscan/Corbis; **p89B:** REX/Sipa Press; **p89C:** UIG via Getty Images; **p92A:** Colin Monteath/ Hedgehog House/Minden Pictures/ Corbis; **p97E:** WaterAid/Josh Hobbins; **p99C (top):** Photofusion Picture Library/Alamy; **p99C (bottom):** Ben Curtis/PA Archive/Press Association Images; **p100C (left and right):** David Waugh; **p101D:** age fotostock/ Robert Harding; **p102A:** Rex Features; **p103D:** David Waugh; **p104A:** Images of Africa Photobank/Alamy; **p112:** michaeljung/Fotolia; **p114:** michaeljung/Fotolia; **p117:** kim/Fotolia; **back cover:** (UK) Rex Features/ Frank Siteman; (USA) Digital Vision PB (NT); (Brazil) Robert Fried/ Alamy; (China) David Waugh; (Kenya) Peter Horee/Alamy; (Bangladesh) Robert Harding Picture Library Ltd/Alamy.

Contents

How is the landscape shaped?

What is this unit about?

This unit explains how our landscape is affected by erosion, transportation and deposition through the work of rivers, the sea and glaciation. It also looks at how coastal erosion can cause severe problems for people.

In this unit you will learn about:

- different types of weathering
- how material is eroded, transported and deposited
- how rivers shape the land
- how the sea shapes the coast
- the problems of coastal erosion
- how coastal erosion may be reduced
- the landforms that result from river, coastal and glaciation processes.

Why is learning about rivers and coasts important?

In Britain we are never very far from a river, coast or glaciated area. Nearby areas:

- are often very attractive
- may be used for recreation purposes
- provide sites for settlements and industries.

For these reasons, we need to understand river, coastal and glaciation processes so that we can make the most of their features and manage any problems that might arise. This unit should also help you develop an interest and appreciation of the landscape and scenery around you.

A Angel Falls, Venezuela

B Grand Canyon, USA

C Florida Keys, USA

- For each of the three photos, make a list of words to describe it.
- How do the photos show the power of rivers and the sea?
- Of the places shown in these photos, which:
  - would you most like to visit
  - do you think is the most spectacular
  - do you think looks the most dangerous
  - would you like to know more about?

Give reasons for your answers.

What is weathering?

There is a great variety of different scenery in the world. Some places are mountainous, some are flat, some can be described as spectacular and others simply as interesting. Geographers call the scenery of a place the **landscape**. Some examples of the world's landscapes are shown in photos **A**, **B** and **C**.

The surface of the earth and the landscapes we see around us not only differ from place to place, but they are changing all the time.

Rain, sun, wind and frost constantly break down the rocks. Great mountain ranges get worn down, valleys are made wider and deeper, and coastlines are changed. The breaking up of the earth's surface in this way is due to **weathering** and **erosion**. Weathering takes place when the rocks are attacked by the weather. Erosion is the wearing away of the land. These two pages show some examples of weathering. Erosion is explained more fully on pages 8 and 9.

A Mount Everest

B Monument Valley, USA

C Guilin, China

Freeze–thaw weathering

This can also be called frost shattering. Water may get into a crack in a rock and freeze. As the water turns to ice it expands and causes the crack to open a little. When it thaws the ice melts and changes back to water. Repeated freezing and thawing weakens the rock and splits it into jagged pieces. This type of weathering is common in mountainous areas where temperatures are often around freezing point.

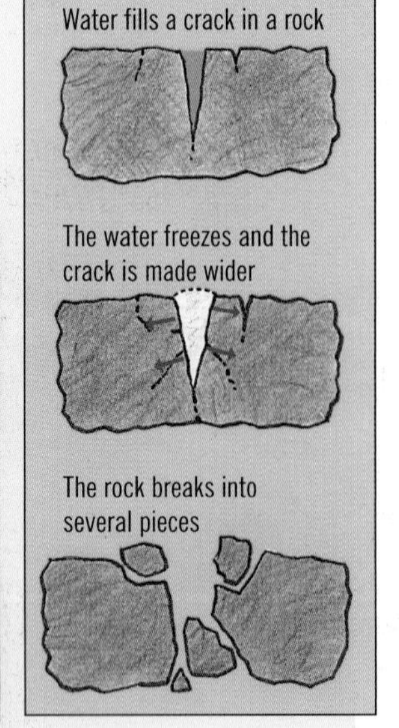

Onion-skin weathering

This happens when a rock is repeatedly heated and cooled. As it is heated, the outer layer of the rock expands slightly and as it cools the rock contracts. Continual expansion and contraction causes small pieces of the rock surface to peel off like the skin of an onion. This type of weathering is common in desert areas where it is very hot during the day but cool at night.

Biological weathering

This is due to the action of plants and animals. Seeds may fall into cracks in the rocks where shelter and moisture help them grow into small plants or trees. As the roots develop they gradually force the cracks to widen and the rock to fall apart. Eventually whole rocks can be broken into small pieces. Burrowing animals such as rabbits, moles and even earthworms can also help break down rock.

Chemical weathering

This is caused by the action of water. Ordinary rainwater contains small amounts of acid. When it comes into contact with rock the acid attacks it and causes the rock to rot and crumble away. The results of this can be seen on buildings and in churchyards where the stone has been worn away or pitted. Water and heat make chemical weathering happen faster, so it is greatest in places that are warm and wet.

Activities

1 Make a larger copy of diagram **D**.
   a Write in the meaning of weathering.
   b Add labels to the weathering features.

D

2 Copy and complete these sentences.
   a Freeze–thaw weathering is ...
   b Onion-skin weathering is ...
   c Chemical weathering is ...

3 With the help of a labelled diagram, show how freeze–thaw weathering can break up rocks.

4 a Make a larger copy of diagram **E**.
   b Show how root action can break up rocks, by adding the labels in **F** to the correct boxes. Give your diagram a title.

E

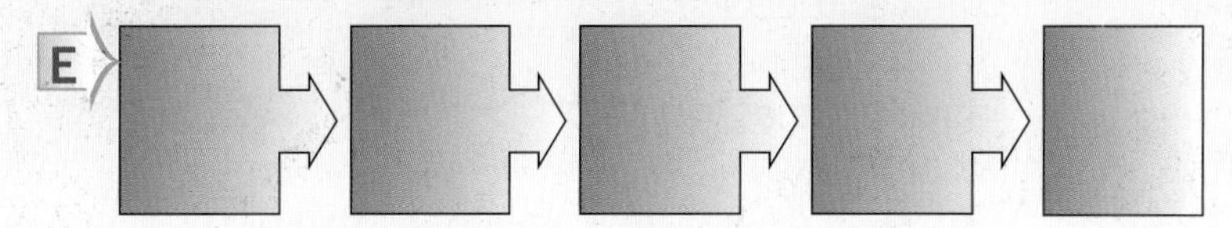

5 a Draw these simple sketches of photos **A**, **B** and **C**.
   b Give each sketch a title and underneath say what type of weathering is likely to be most important there. Give reasons for your answer.

F

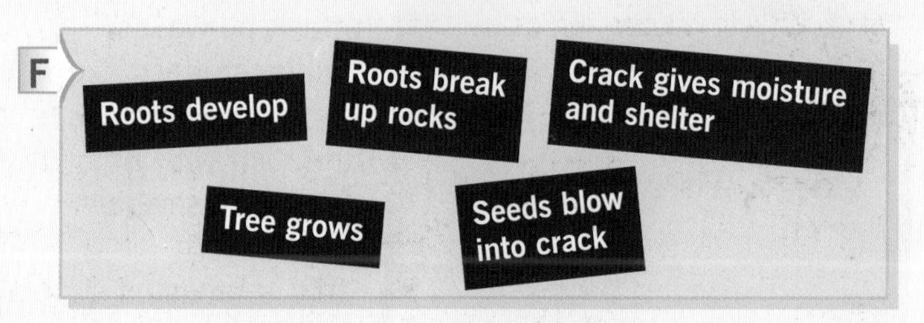

Summary

Weathering is the breakdown of rocks by water, frost and temperature change. Rocks can also be broken down by the effects of plants and animals.

What is erosion …

Weathering and erosion work together. Weathering breaks up and weakens the surface of rocks while erosion wears away and removes the loosened material. The action of rivers, the sea, ice and wind are the chief types of erosion. Human erosion is also important. Bulldozers and lorries can dig out and move large amounts of soil and loose rock, so changing the landscape. People also remove trees and vegetation which can allow water, wind and ice to erode land more easily.

The work of rivers, the sea, ice and wind are explained in **A** below.

A

Rivers

Every day rivers wear away tiny bits of rock from their bed, and eat into the banks on either side of the channel. This material is carried downstream and deposited when the water slows down. In times of flood, large boulders may be loosened and rolled down the river bed.

Ice

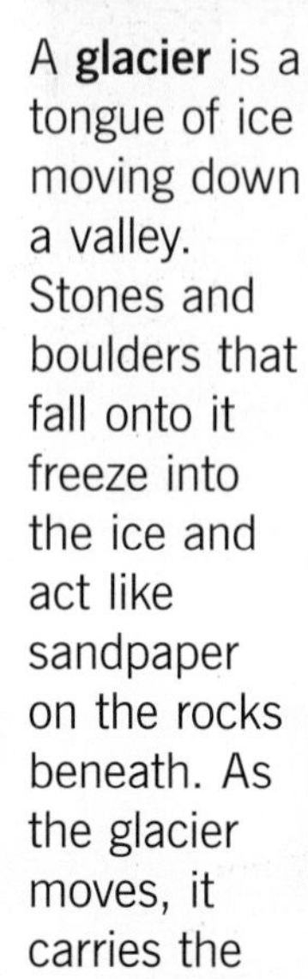

A **glacier** is a tongue of ice moving down a valley. Stones and boulders that fall onto it freeze into the ice and act like sandpaper on the rocks beneath. As the glacier moves, it carries the material downwards and at the same time wears away the valley bottom and sides.

Sea

Coastlines are under constant attack by **waves**. During storms each wave hits the rock with a weight of several tonnes. When this is repeated many times, the rock is weakened and pieces break off. **Currents** carry loose material away and deposit it elsewhere.

Wind

Explorers who cross deserts in cars often find their paintwork worn away and their windscreens scratched. This is because the wind picks up tiny particles of sand and blasts them against anything that is in the way. Rocks in desert areas are often eroded into strange shapes by this sandblasting effect.

... and how can it help shape the land?

Look at cartoon B on the right. It shows some gardeners who are trying to alter a garden by digging out soil (erosion), moving it in a wheelbarrow (transportation), and dumping it somewhere else (deposition). The more energy they have, the more soil they can dig or transport. When they are tired the digging slows down and they lack the strength to push the barrow, resulting in it toppling over and dumping its load.

On a larger scale, mountains, valleys, plains and coasts are shaped and changed by water, ice and wind. **Erosion** wears away the land, **transportation** moves the material from one place to another, and **deposition** builds up new landforms.

B

Activities

1 a List the following in order of how hard they are. Give the hardest first and the softest last.

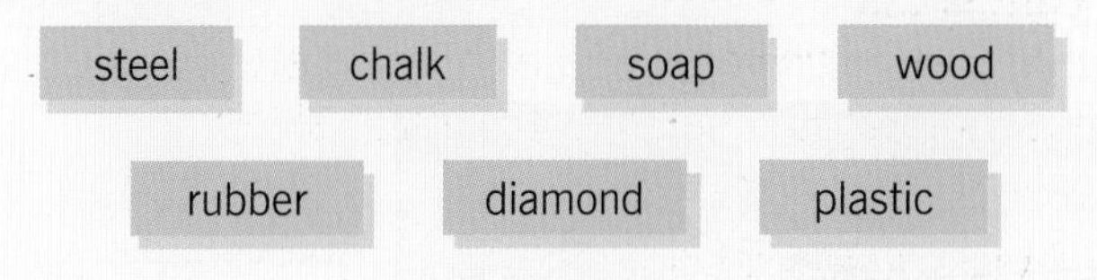

b Put a line under the two you think would be the most difficult to wear down.

c Choose any three from your list and say how they might be worn down.

2 Of the five statements below, three are correct. Write out the correct ones.

- Weathering is the breakdown of rock by nature.
- Erosion is the wearing away of rock.
- Weathering and erosion are the same.
- Weathering moves material from one place to another.
- Erosion includes the removal of loose material.

3 a Make a large copy of table C.

b Add labels to each drawing.

c Write a short description for each type of erosion.

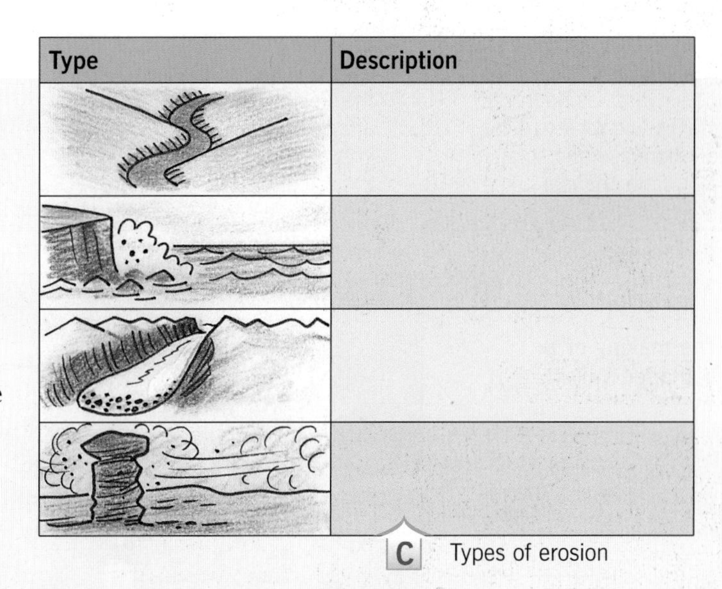

Type	Description

C Types of erosion

4 Cartoon B shows erosion, transportation and deposition in a garden. How else could this be shown? What about a bulldozer, washing dishes or sandpapering wood? For one of these ideas, or for one of your own, draw a simple labelled cartoon to show how it works.

Summary

Erosion is the wearing away of rock and its removal by streams, ice, waves and wind. Erosion, transportation and deposition help shape the land.

How do rivers shape the land?

Rivers work hard. They hardly stop and they continually erode and move material downstream. They are a major force in shaping and altering the land. Running water by itself actually has little power to wear away rocks. What happens is that the water pushes boulders, stones and rock particles along the river's course. As it does so, the loose material scrapes the river bed and banks and loosens other material. Much of what is worn away is then transported by the river and put down somewhere else. In this way rivers can wear out and deepen valleys. They can also change their shape by depositing material.

The landforms to be seen along a river change as it flows from source to mouth. These two pages explain the features of a river in its upper course which is usually in the hills or mountains. Diagram **A** and photo **B** show how a river cuts out a steep-sided valley that is V-shaped.

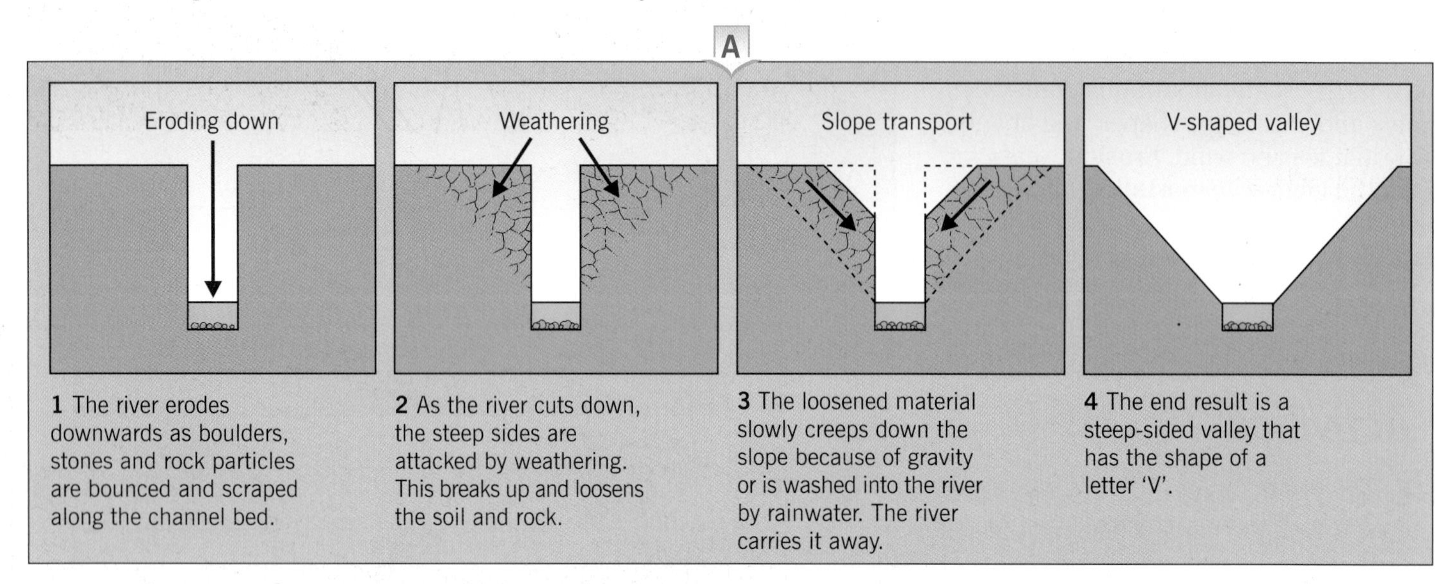

Table **C** below and sketch **D** give some features of a river and its valley.

C

Source	Where a river starts
Spurs	Ridges of land around which a river winds
Valley sides	The slopes on either side of a river
V-shaped valley	The shape of a valley in its upper course
Channel	The course of a river
River banks	The sides of a river channel
River bed	The bottom of a river channel
Load	Material that is carried or moved by the river

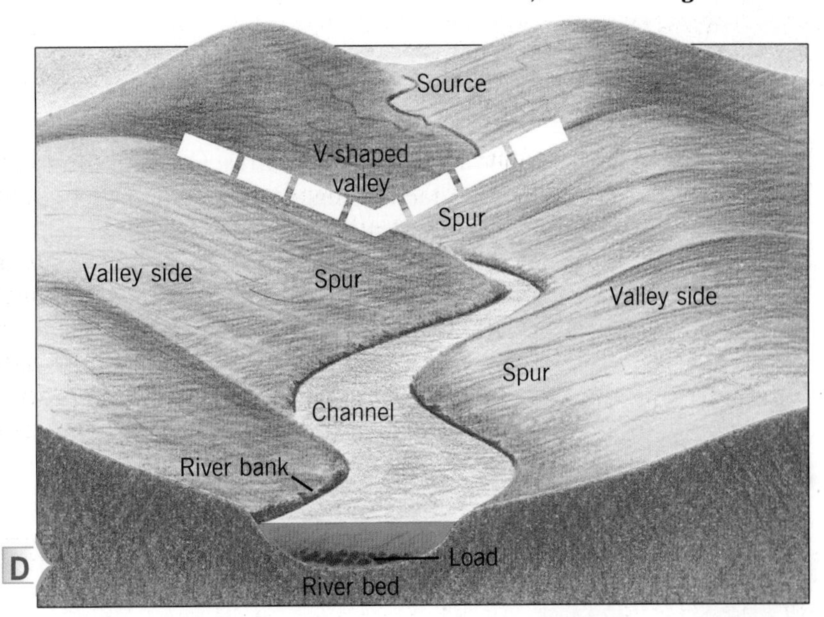

D

Activities

1 Describe how rivers erode their channels. Include these words in your description:

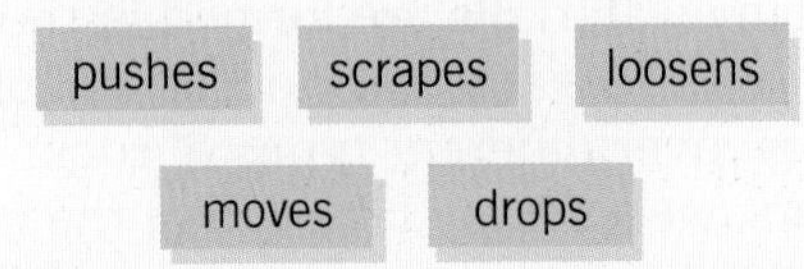

2 a Make a large copy of diagram **E**.

b Show how a valley gets to be V-shaped by describing what happens at ①, ②, ③ and ④.

c Give your diagram a suitable title.

E

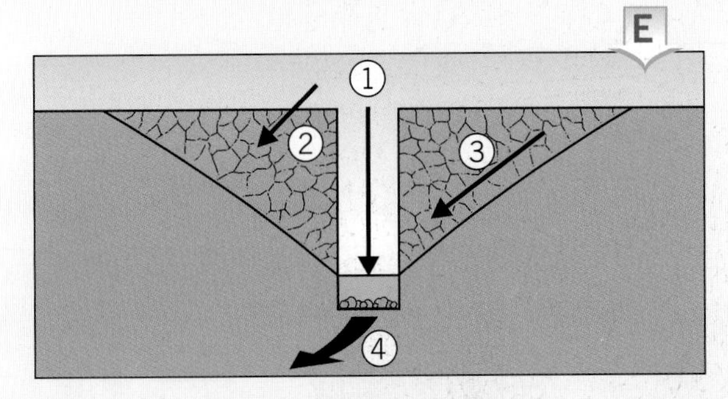

3 a Sketch **F** is a simplified drawing of the river valley shown in photo **B**. Make a copy of the sketch.

b Add the terms below to your sketch in the correct places. The information at the top of this page will help you.

F

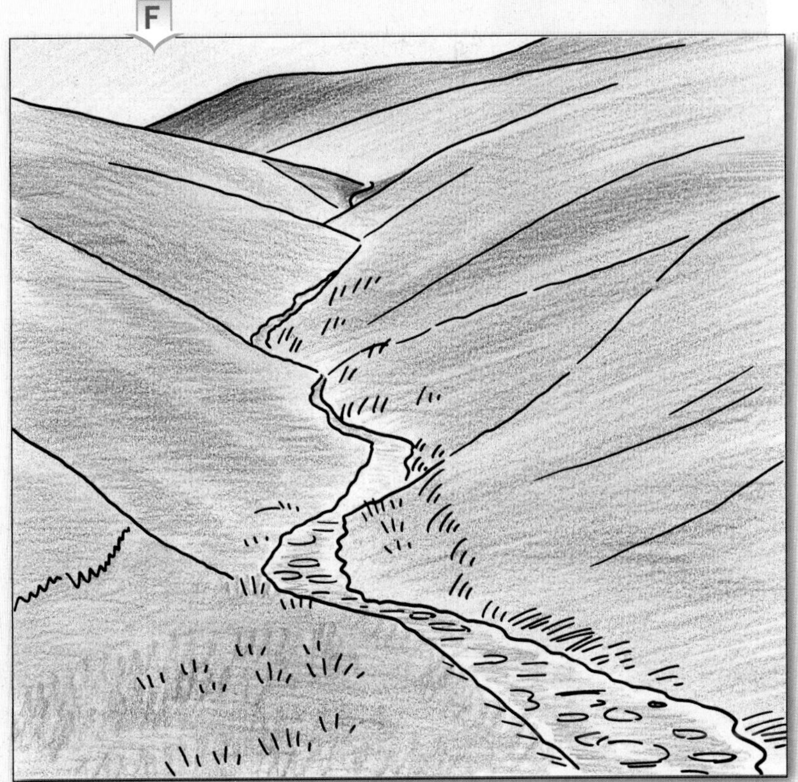

Summary

Rivers erode, transport and deposit material. This helps shape the land. V-shaped valleys are a common feature of a river in its upper course.

What causes waterfalls?

A

Waterfalls are an attractive and often spectacular feature of a river. The highest waterfall in the world is the Angel Falls in South America. Its total height is 979 metres. That is about four times the height of One Canada Square at Canary Wharf in London's Docklands. Waterfalls in Britain are much smaller than this (diagram **A**). One of the finest is High Force in the north of England. It has a height of just 20 metres. It is most impressive in times of flood.

Probably the best-known waterfall in the world is Niagara Falls. It lies on the Niagara River which forms part of the border between Canada and the United States.

B

Lake Erie
CANADA
Niagara River
Goat Island
Horseshoe Falls
UNITED STATES OF AMERICA
American Falls
Original position of falls
Niagara Gorge
11 km
CANADA
N
Hard rock
Soft rock

In this area, a band of hard limestone rock lies on top of softer shales and sandstone. The river flows over the top of the hard rock then plunges down a 50 metre cliff. At the bottom of the cliff the water has worn away the softer rocks to form a pool over 50 metres deep. This is called a **plunge pool**. Down from the falls is the Niagara Gorge. A **gorge** is a valley with almost vertical sides that has been carved out by the river and the waterfall. Photo **D** shows the gorge and waterfall at Niagara.

Sketch **B** shows the Niagara Falls area. The falls here are eating into the cliffs behind the waterfall at nearly one metre a year. The gorge that has been left behind is now 11 kilometres long.

Many waterfalls are formed in the same way as Niagara. They occur when rivers flow over different types of rock. The soft rock wears away faster than the hard rock. In time a step develops over which the river plunges as a waterfall. Water also cuts away rock behind the waterfall. This causes the falls to move back and leave a gorge as it goes. Diagram **C** shows how a waterfall may be worn away by a river.

C

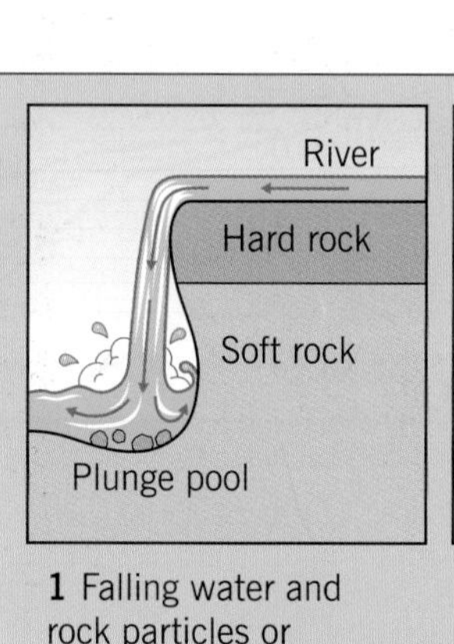

1 Falling water and rock particles or boulders loosen and wear away the softer rock.

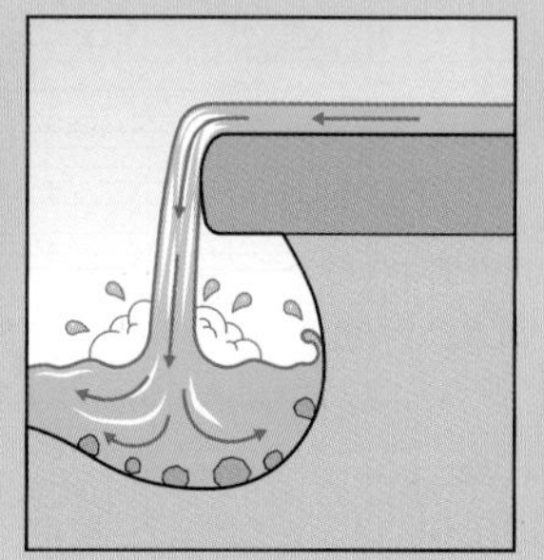

2 The hard rock above is undercut as erosion of the soft rock continues.

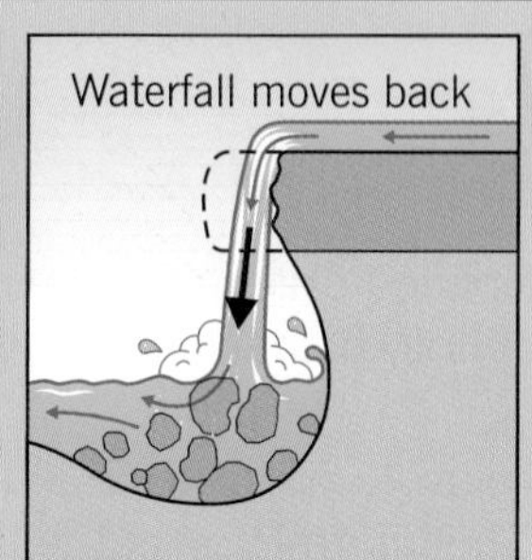

3 The hard rock collapses into the plunge pool to be broken up and washed away by the river. The position of the falls moves back.

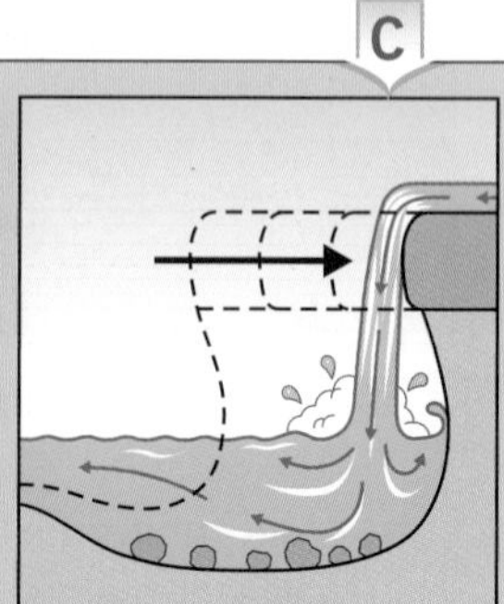

4 Erosion continues and the waterfall slowly eats its way upstream, leaving a gorge behind.

D Niagara Falls, USA

Activities

1 Map **E** shows the Niagara Falls area.
   a Make an accurate copy of the map.
   b Colour the water blue and the land area green.
   c Label the following:

USA | Canada | American Falls | Niagara River | Horseshoe Falls | Niagara Gorge | Goat Island

   d Draw on and label the original position of the falls.
   e The falls have taken 30,000 years to wear back 11 km. Draw on and label where the falls might be 10,000 years from now.

E

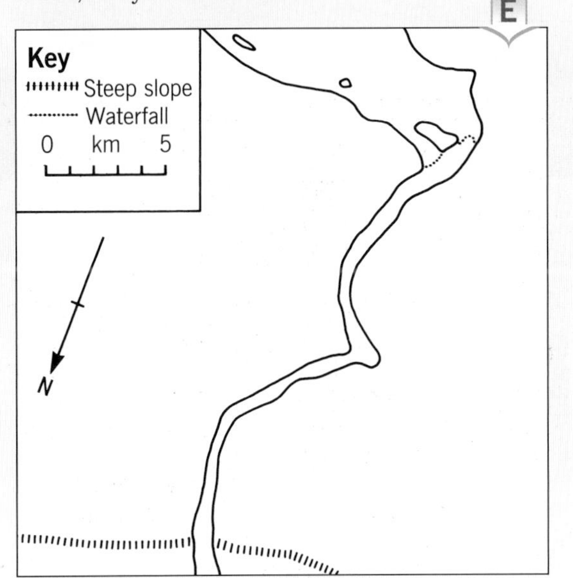

2 a Make a larger copy of diagram **F**.
   b Put these labels in the correct places.

Hard rock | Soft rock | Plunge pool | Hard rock breaks off | Eroded material | Undercutting | Waterfall moves back

F

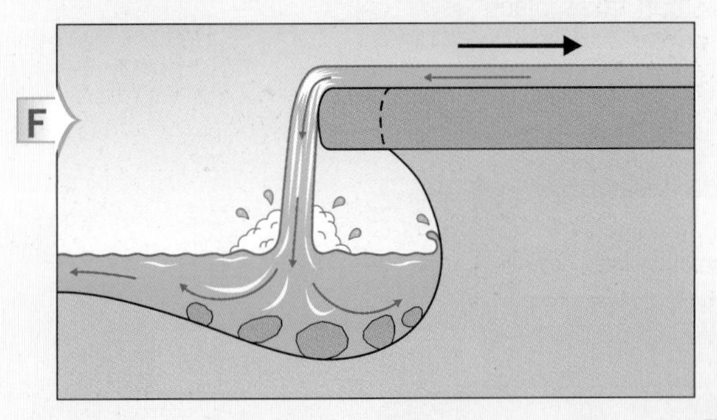

3 Sort the phrases in **G** into the correct order and link them with arrows to show how a waterfall may be worn away by a river.

G

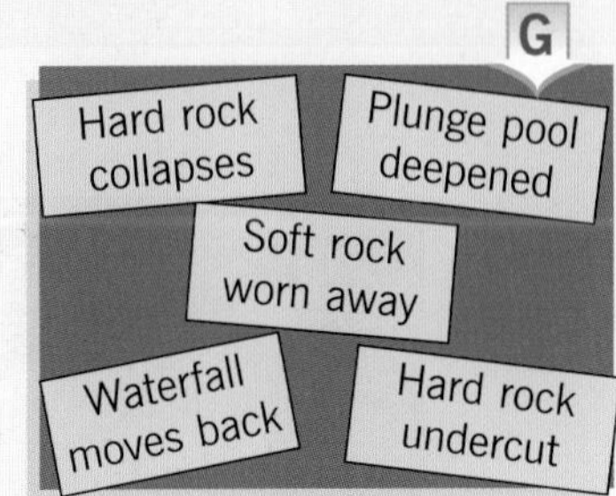

Summary

Many waterfalls are a result of water wearing away soft rock more quickly than the hard rock. As a waterfall erodes back, a gorge may be produced.

What happens on a river bend?

Have you noticed that rivers rarely flow in a straight line? Usually they twist and turn as they make their way down to the sea. The only time they are straight seems to be when people interfere with them by building banks or diverting their course.

Bends develop on a river mainly because of the water's eroding power. Think about when you are a passenger in a car and it goes around a corner. You are thrown towards the outside of the curve, often with quite a lot of force. The same happens when a river goes around a bend. The force of the water is greatest towards the outside of the bend. When it hits the bank it causes erosion. This erosion deepens the channel at that point and wears away the bank to make a small **river cliff**. On the inside of the bend, the water movement, or **current**, is slower. Material builds up here due to deposition. This makes the bank gently sloping and the river channel shallow.

Diagram **A** shows what happens on a river bend. At the bottom of the diagram is a **cross-section**. This shows what the river would look like if a slice was cut across it from one side to the other.

A

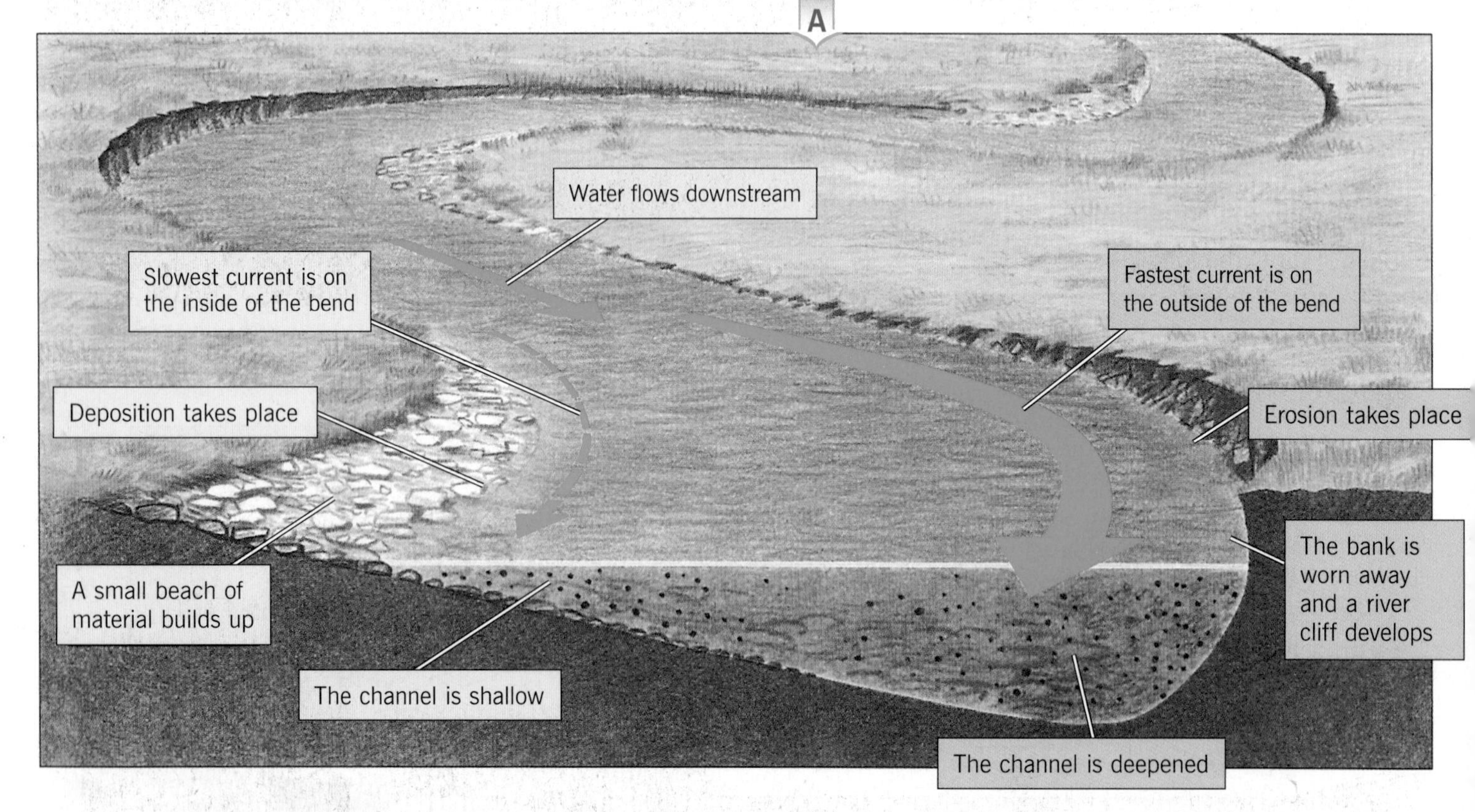

B Deposition on the inside of a river bend

C Erosion on the outside of a river bend

Look at sketch **D** and photo **E**. The river has many bends. These are called **meanders** and are a common feature of most rivers. On either side of the river channel there is an area of flat land called the **flood plain**. This area gets covered in water when the river overflows its banks. Flood plains are made up of **alluvium**, a fine muddy material that is left behind after floods. Alluvium is sometimes called **silt**.

Flood plains are useful to people because they are areas of flat land and have rich fertile soil. This makes them good for building on and for farming.

D

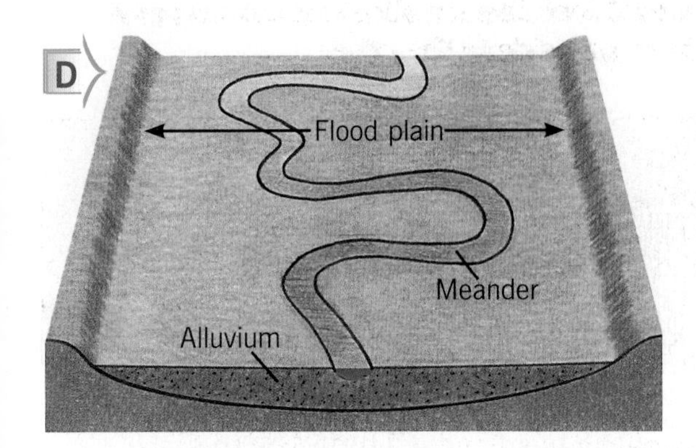

E

Activities

1. Look at diagram **F**, which is a simple cross-section of a river bend.
   - **a** Draw the cross-section.
   - **b** Write the labels from **G** in the correct places.
   - **c** Give your drawing a title.
   - **d** Describe why one side of the river bend is different from the other.

F

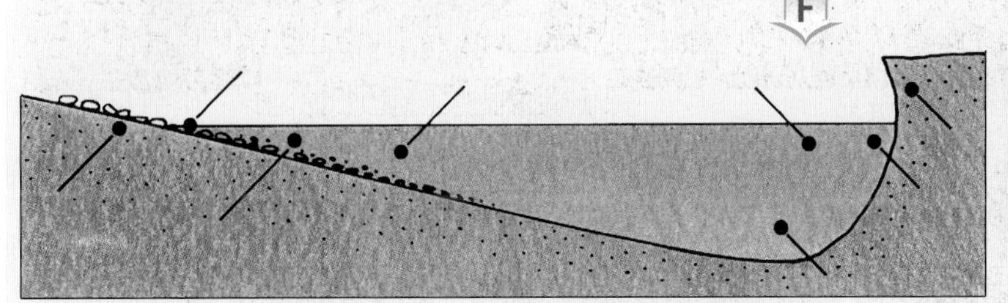

2.
   - **a** Make simple sketches of photos **B** and **C**.
   - **b** For each sketch describe the river feature that it shows.
   - **c** Explain how each feature was made.
3. Give the meaning of the terms shown in sketch **D**.
4. Write down two reasons why the flood plain of a valley is good for farming.
5. Give one problem of farming the flood plain. Suggest what could be done to reduce the problem.

G

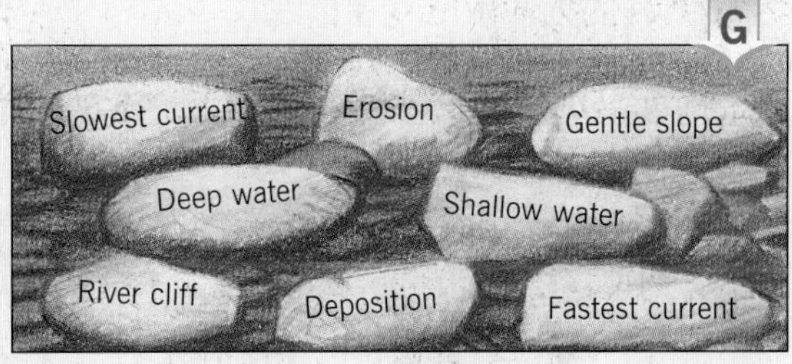

Summary

A river course is seldom straight. It usually has many bends which cause it to meander down its valley. The outside of a river bend is worn away by erosion while the inside is built up by deposition.

How does the sea shape the coast?

The sea is never still. On quiet days the movement is slow and gentle, and the sea is flat and almost calm. On stormy days large waves crash against the shore. These large waves have such force that they can drive a ship against the rocks or smash up sea defences and piers. The sea can also wear away the coast and move bits of rock and sand from one place to another. This ability to erode, transport and deposit material produces many interesting coastal landforms.

Erosion landforms are made by the wearing away of the coast (photo A). In stormy conditions the sea picks up loose rocks and throws them at the shore. This bombardment undercuts **cliffs**, opens up cracks and breaks up loose rocks into smaller and smaller pieces. Areas which have soft rocks are worn away more easily than those with hard rocks. The soft rock areas become **bays** and the hard rock areas become **headlands**. A bay is an opening in the coastline. A headland is a stretch of land jutting out into the sea.

A Old Harry Rocks, Dorset, UK

Sketch B shows how a headland is eroded by the sea and how other landforms develop.

B

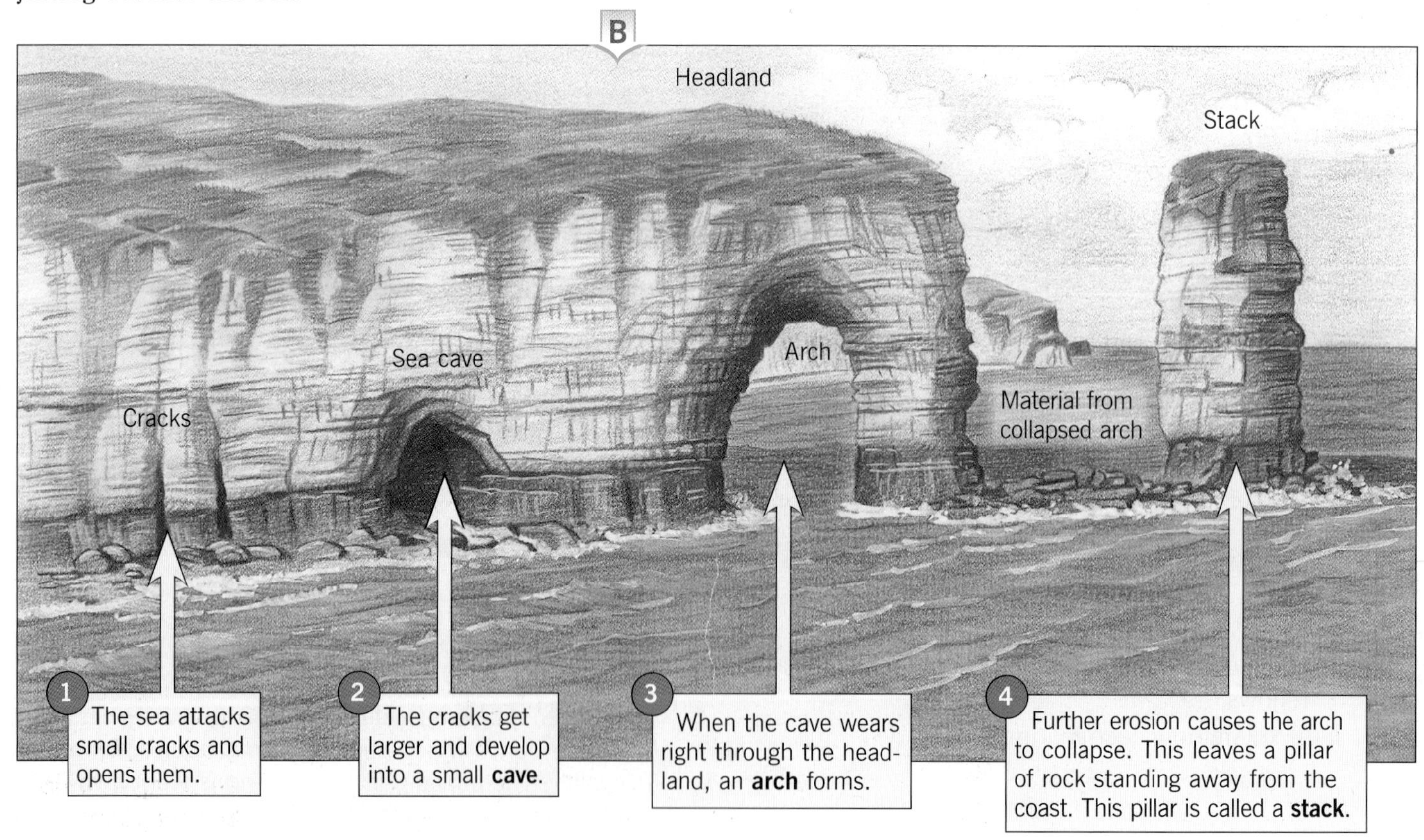

Beaches are one of the most common features of our shoreline. They are formed when material worn away from one part of the coast is carried along and dropped somewhere else. A beach is an example of a **deposition landform**.

A **spit** is a special type of beach extending out into the sea. It is a long finger of sand and shingle that often grows out across a bay or the mouth of a river.

Photo **C** and drawing **D** show Spurn Head spit. The spit is 6 kilometres long and forms a sweeping curve that stretches halfway across the mouth of the River Humber. It is continually changing its shape as new material is deposited and old material is worn away or moved elsewhere.

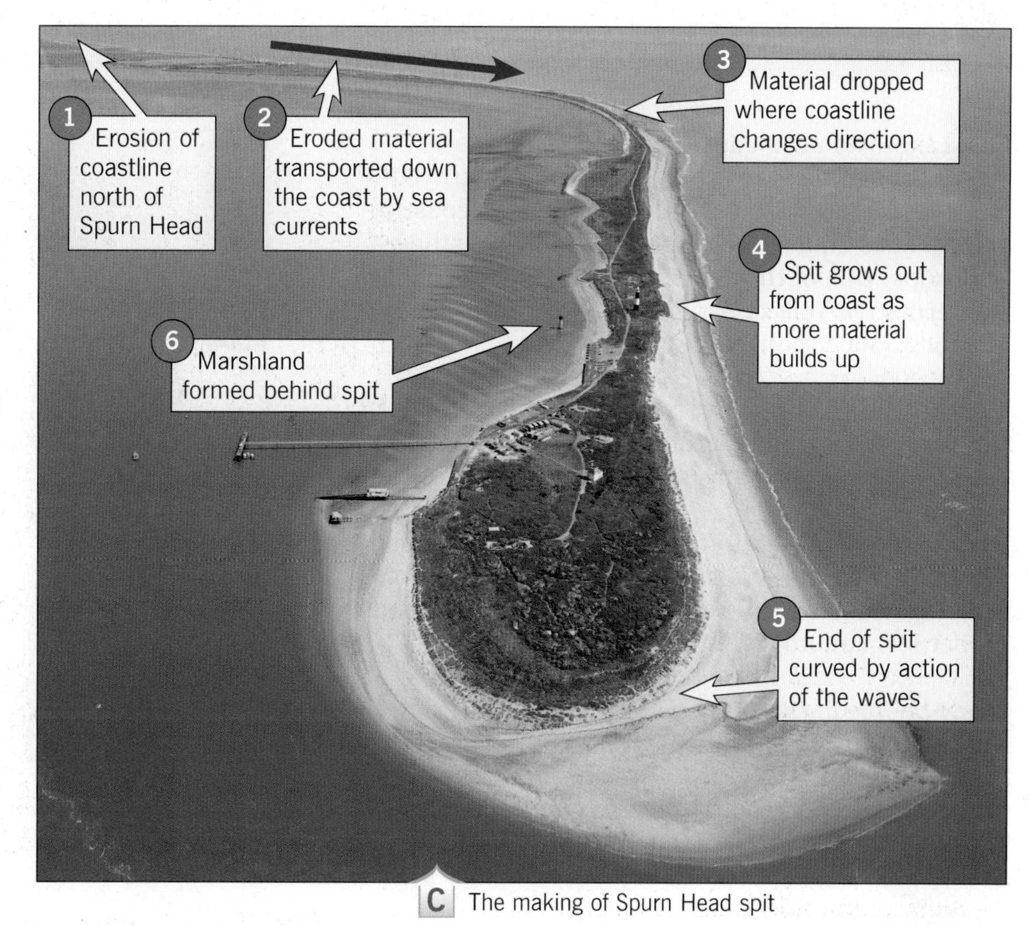

C The making of Spurn Head spit

Activities

1 a Make a sketch of photo **A**.
  b Label these features on your sketch:

crack | arch | cave | stack | material from a collapsed arch

  c Explain how the arch was formed.
  d Draw a dotted line to where there was once another arch.

2 a Copy the drawings of Spurn Head spit shown below (**D**).
  b With the help of the drawings, describe the formation of the spit. You could add information to the drawings as well as writing an explanation underneath. Include these terms in your description.

erosion | Flamborough Head | Spurn Head | 6 km | currents | deposition | transportation | grows

D

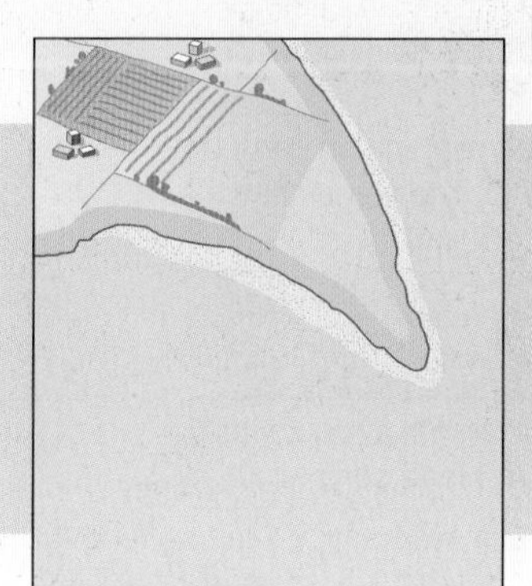

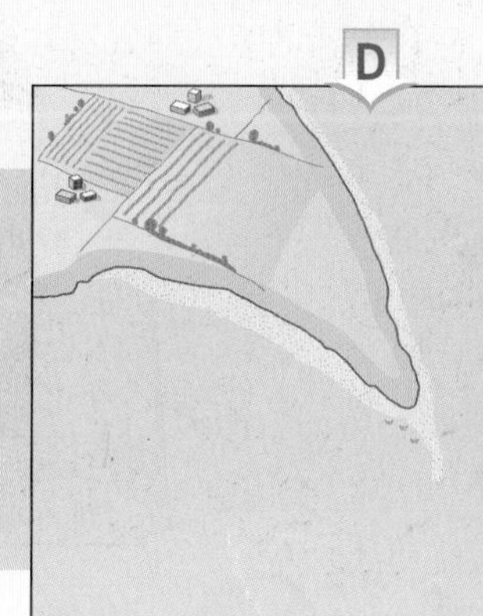

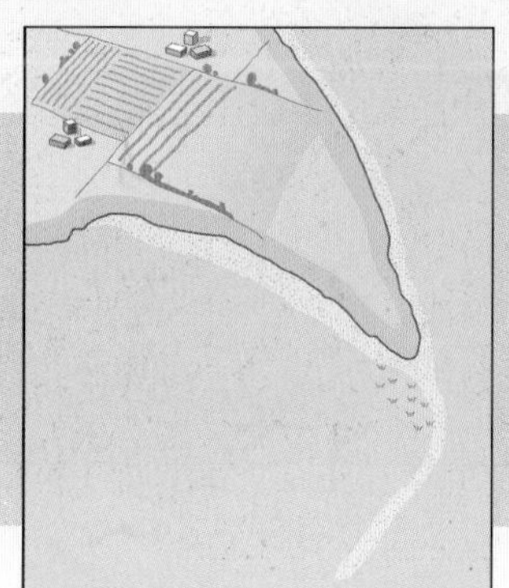

Summary

The coastline is always changing its shape. Some parts are being worn away by erosion while other parts are being built up by deposition.

What is the coastal erosion problem?

Coastal erosion can cause severe problems. Agricultural land may be lost, buildings destroyed and transport links put in danger. The east coast of England has some of the fastest-eroding coastlines in Europe. One of the areas in most danger is Holderness in Yorkshire where the sea is eroding the land at about 2 metres every year.

Over the last 2,000 years, several villages and farms have disappeared into the North Sea as the Holderness coastline has gradually moved back.

The village of Mappleton is the latest victim. Already several houses have fallen into the sea, some farm buildings have become unsafe and the coastal road is all but lost (photo **A**).

Many people in the Holderness area are worried about their future. They are afraid of losing their homes and, in some cases, their livelihoods. Several farms are threatened and seaside resorts like Hornsea and Withernsea, where many people work, are also in danger. Two cliff-top gas plants for gas piped from the North Sea are also at risk.

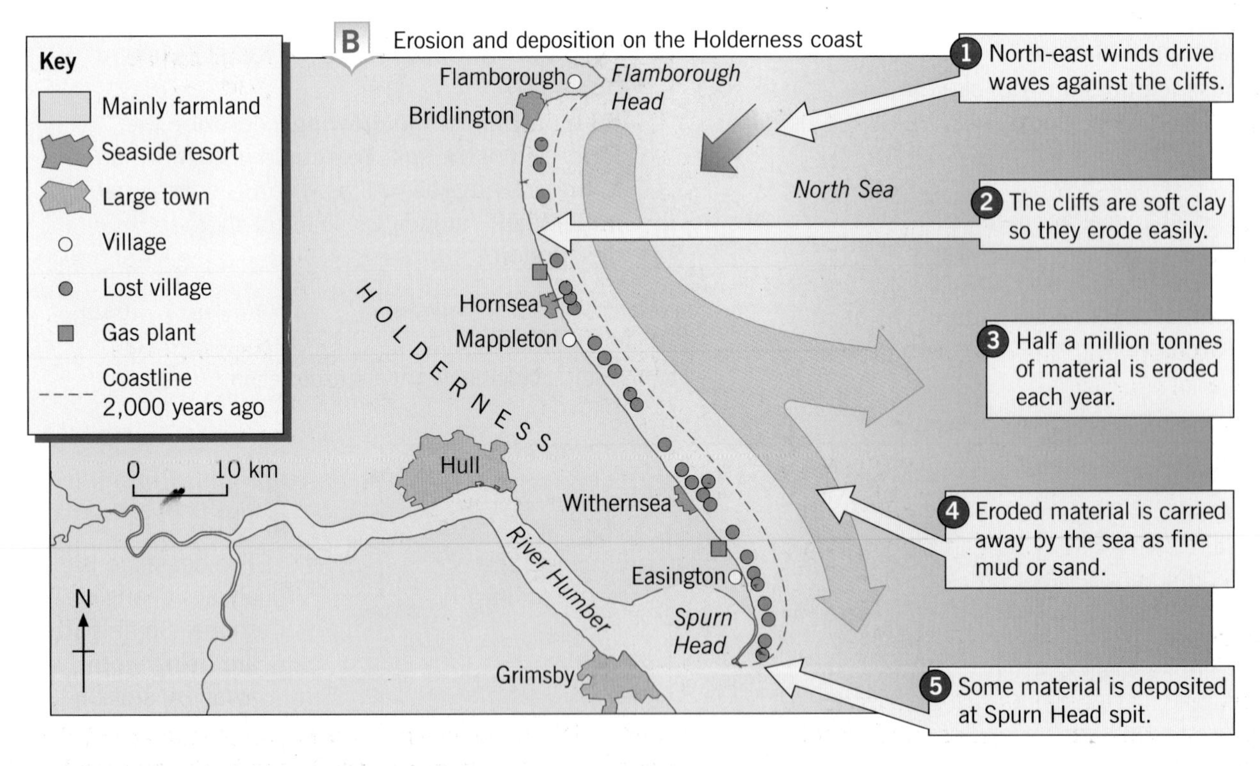

All around Britain's coast, the cliffs are being worn away by waves and weathering. It happens all the time. The process never stops. But why is erosion such a problem along the Holderness coast and why is it worse here than on any other part of the coastline?

The main reason is that the rock in this area is soft boulder clay left by the glaciers during the last Ice Age some 10,000 years ago. It is easily worn away by weathering and the constant pounding of waves. Drawing **C** shows the processes at work along the Holderness coast.

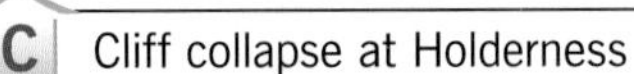
C Cliff collapse at Holderness

Activities

1 Look at map **B**.
   a How far has the coastline moved back in 2,000 years?
   b How many villages have been lost?
   c Suggest why the coast near Grimsby has not been affected by erosion.

2 Look at maps **D** and **E**.
   a How far has the coastline at Mappleton moved back over 30 years?
   b How many buildings have been lost?
   c Which buildings should soon be closed down?

3 Make a list of the problems caused by coastal erosion in the Holderness area. Sort the problems under the headings:

   Buildings | Industries | Employment

4 Many people like Mappleton as a place to live. Suggest three reasons **for** living there and three reasons **against**.

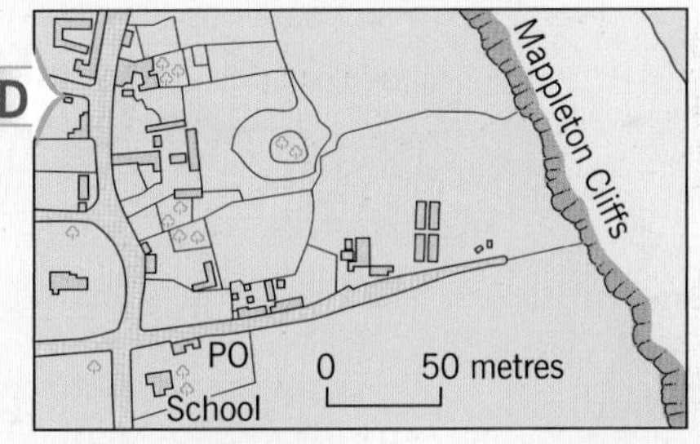

D Mappleton 30 years ago

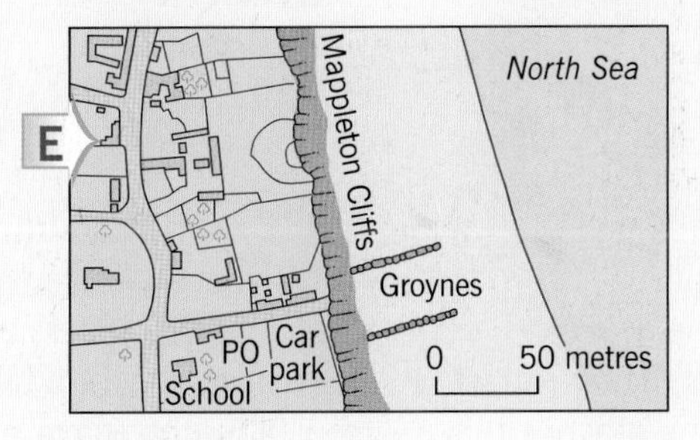

E Mappleton today

Summary

Erosion is a problem for many of our coastal areas. It causes land loss and may destroy property, transport links and industries. It can also result in job losses.

How can coastal erosion be reduced?

Protecting coasts can be both difficult and expensive. Where valuable land or property is under threat from the sea, local authorities try to slow down or prevent erosion. Since 2000, over £1 billion has been spent defending Britain's coastline. Photo **A** and drawing **B** show some of the methods that can be used.

Unfortunately, putting in new sea defences is not always the best solution. Geographers know that protecting one part of the coast can cause even worse problems further along the same coastline.

At Mappleton, for example, new sea defences built in 1992 have helped protect the village but have led to greater erosion of the cliffs to the south.

Many people nowadays are actually against building coastal defences. They support schemes which work with nature rather than against it. Schemes like this, they say, do less damage and help retain wildlife and the quality of the natural environment.

A Sea defences

B Some methods used to help reduce coastal erosion

Sea walls
Sea walls stop the waves reaching the land. They reflect the waves back to sea but this can wash away the beach. They give good protection but are expensive and may need to be repaired in time.
Cost: *about £7 million per km*

Beach rebuilding
This replaces the sand and shingle which has been lost from the beach. The beach absorbs wave energy and is a good defence against the sea. It protects the land or sea wall behind the beach and looks more natural.
Cost: *about £2 million per km*

Groynes
These are built down the beach and into the sea about 200 metres apart. They slow the movement of material along the coast and help build up the beach. The beach then helps protect the land. Rock groynes are expensive.
Cost: *about £1.5 million per groyne*

The Holderness coast

So what should be done with the Holderness coast? There is no easy solution but what experts do agree on is that any plan must look at managing the whole coastline rather than just parts of it. The following are three possible solutions.

1 Build sea defences along the whole 60 kilometres of coastline. This would be expensive but would be the choice of most local residents who argue that everyone along the coast deserves to be protected. One problem could be that it would cut off the supply of sand to Spurn Head, which might disappear altogether.
2 Protect the main towns and allow the sea to erode the land between these places into small bays, as shown on map **C**. Beaches would form in the bays which would help protect the coast from further erosion. Some material would continue to be transported down to Spurn Head and beyond.
3 A third solution might be to do nothing at all and let nature take its course.

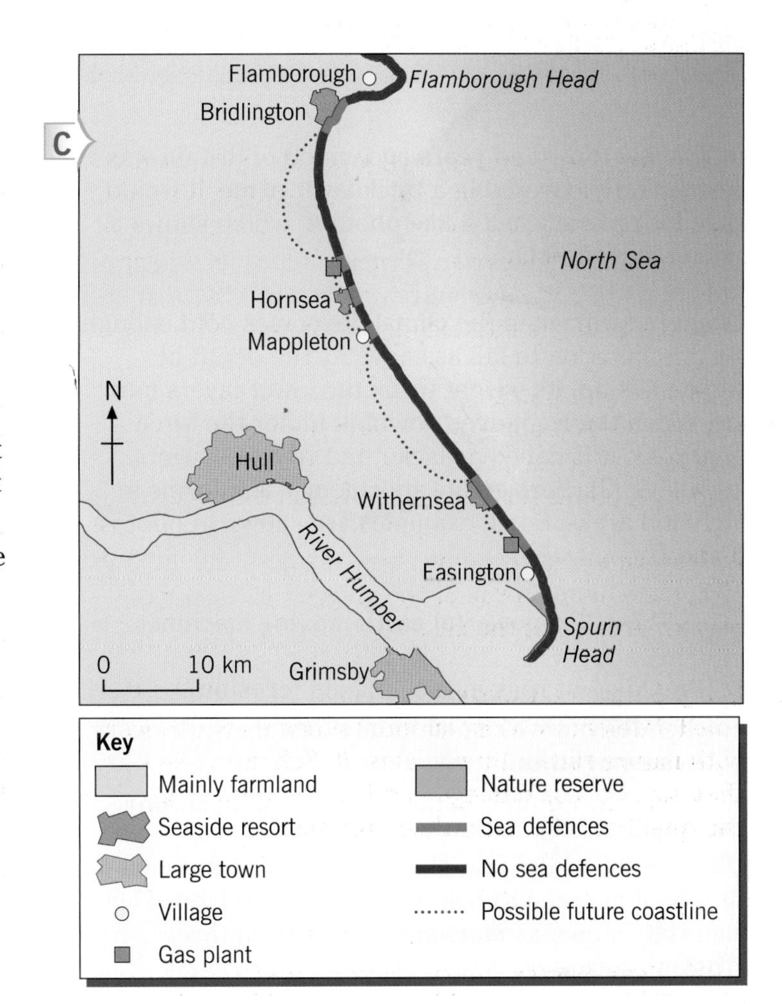

Rip-rap
This is a mixture of large boulders and concrete blocks which protect the coast by breaking up the waves. They don't protect cliffs as well as a sea wall but they do help retain the beach. They can look ugly and make beach access difficult.
Cost: *about £3 million per km*

Activities

1 Look at the coastal defences in drawing **B**.
   a Which do you think would:
      - be easiest to build
      - be most attractive to look at
      - cause fewest problems for people using the beach?
   b Work out the cost of protecting all 60 kilometres of the Holderness coast for each of the methods shown.
2 What are the arguments against protecting the whole coastline?
3 a Which of the three solutions would you choose for Holderness?
   b Describe the scheme.
   c Give the advantages and disadvantages of the scheme.

Summary

Protecting coasts is not easy. There are arguments for and against trying to protect the coastline from erosion.

How does ice shape the land?

Only a few thousand years ago, much of Britain was permanently covered in a thick layer of ice. It would have looked very much like photo A, which shows a glacier in Switzerland.

Glaciers form when the climate becomes cold enough for precipitation to fall as snow. As the depth of snow piles up, its weight turns the lower layers into ice. When the ice moves downhill under the force of gravity, it is called a glacier and replaces rivers in valleys. Glaciers create spectacular landforms in highland areas. Some examples are shown in photos B and D.

Glaciers are like powerful earth-moving machines:

- they dig out rock in the uplands (**erosion**)
- then they move the material down the valleys (**transportation**)
- then they dump it in the lowlands (**deposition**).

Like rivers and the sea, glaciers need a continuous supply of material to help them erode the land. This material, known as **moraine**, results from three erosion processes:

- **freeze–thaw weathering** when rock is broken up by water in cracks freezing and thawing (this is explained on page 6)
- **abrasion** which wears away the rock
- **plucking** where ice freezes onto rock and pulls some of it away as the glacier moves.

A The Gorner glacier in the Swiss Alps

B A corrie (cwm) in Wales, UK. Cadair Idris in winter

C Formation of a corrie

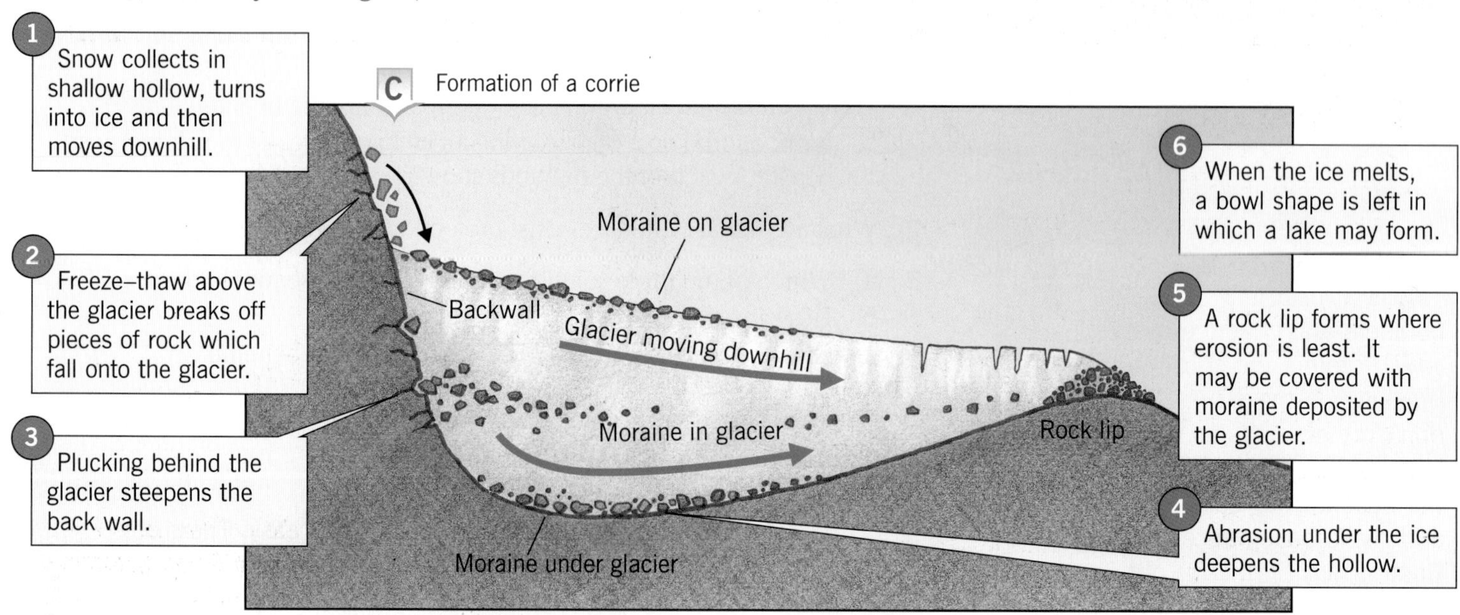

Diagram C shows how **glaciation** can create a **corrie**. Corries are deep, rounded hollows with a steep backwall and sides. They are shaped rather like an armchair. Many corries contain a lake or tarn.

Corries often develop on more than one side of a mountain. When this happens, the land between them gets narrower, due to erosion, until a knife-edged ridge is formed. This is called an **arête**. When three or more corries cut into the same mountain, a **pyramidal peak** is formed. Several arêtes may radiate from the peak. Diagram E shows this process.

D The Matterhorn, Switzerland

E Formation of arêtes and a pyramidal peak

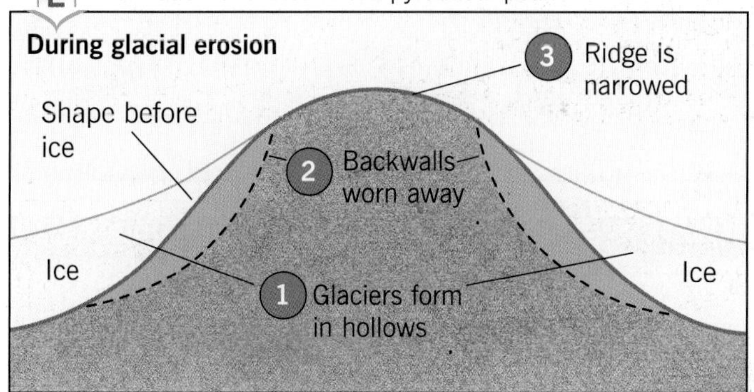

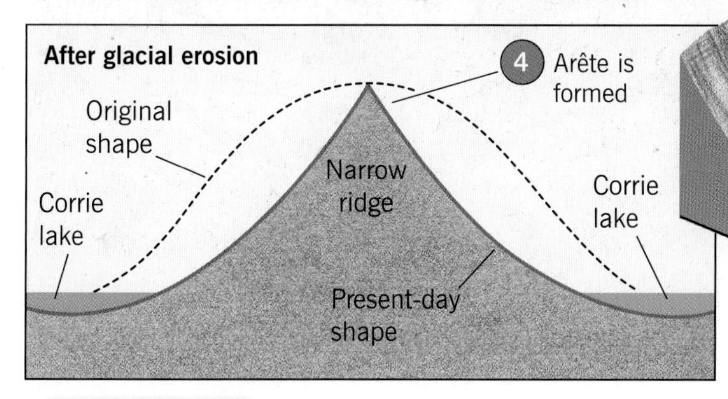

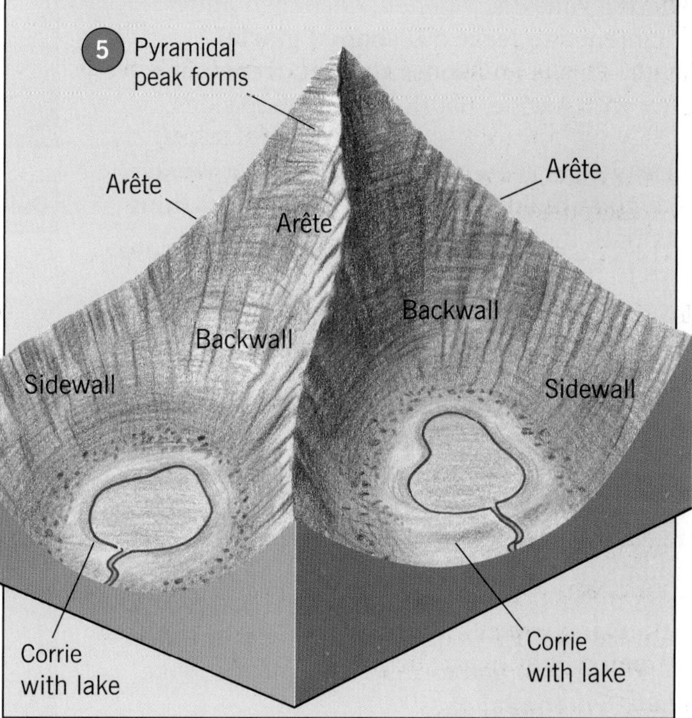

Activities

1 a What is a glacier?
  b What causes a glacier to move downhill?
  c What is moraine?

2 a Make a larger copy of drawing F.
  b Draw in any areas of moraine.
  c Label and describe the erosion processes likely to occur at A, B and C.

3 a Make a sketch of the Matterhorn in photo D.
  b Label the main features.
  c Describe how the Matterhorn formed.

F

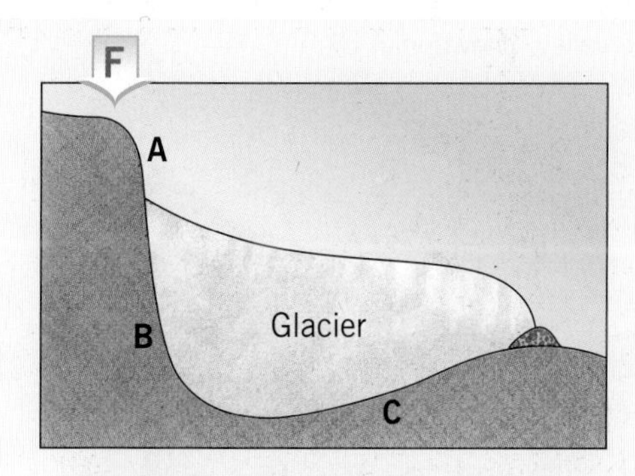

Summary

Many features in our landscape are the action of ice shaping the land. Corries, arêtes and pyramidal peaks are examples of glacial erosion.

What landforms result from glaciation?

Can you remember the shape of a river valley? You will see on pages 10 and 11 that it is **V-shaped**. The top part of drawing **A** shows that same feature. Notice that the river winds around areas of higher land known as interlocking spurs.

Glaciers act differently. They tend to follow the easiest route, which is usually a river valley. Unlike a river valley, however, they fill the entire valley and their power to erode is much greater. The result is that the previous V-shaped valley is widened, deepened and straightened to leave a **U-shaped glacial trough**. This is an almost straight, trench-like valley with a wide, flat floor and steep sides.

As the glacier moves down the valley it works like a giant bulldozer. It wears away everything in its path including the ends of interlocking spurs to leave cliff-like **truncated spurs**.

Glaciers in the smaller tributary valleys have less power than the one in the main valley so erode less quickly. When the ice melts, they are left as **hanging valleys**. The river descends into the main valley as a waterfall.

The glacier in the main valley carries vast amounts of moraine with it. When the ice melts, this moraine is deposited across the valley and a dam-like feature results. Meltwater is trapped behind the dam and forms a long, narrow lake called a **ribbon lake**.

A

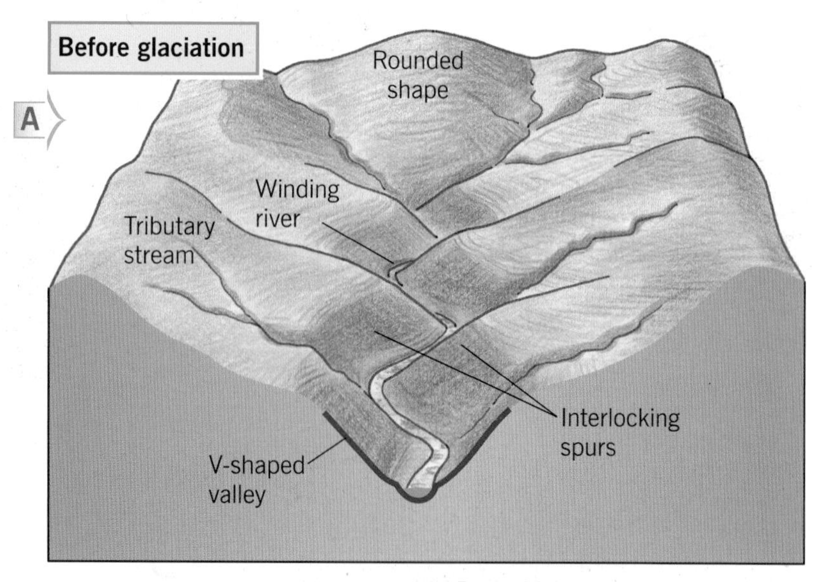

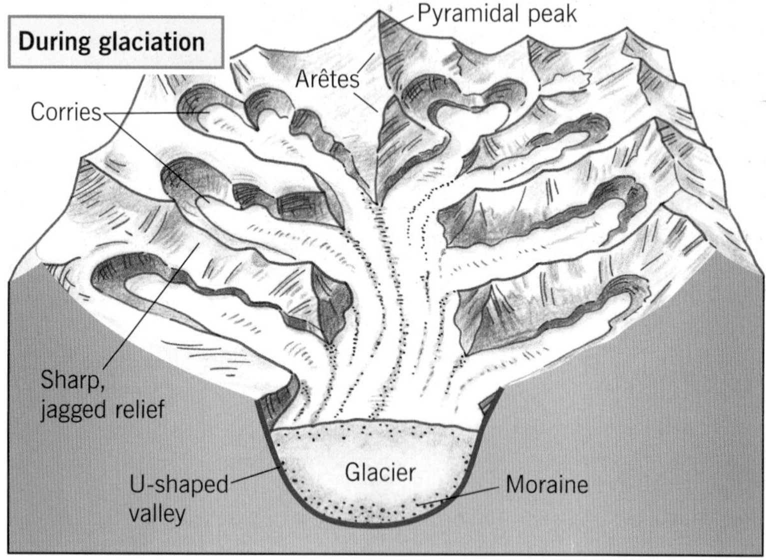

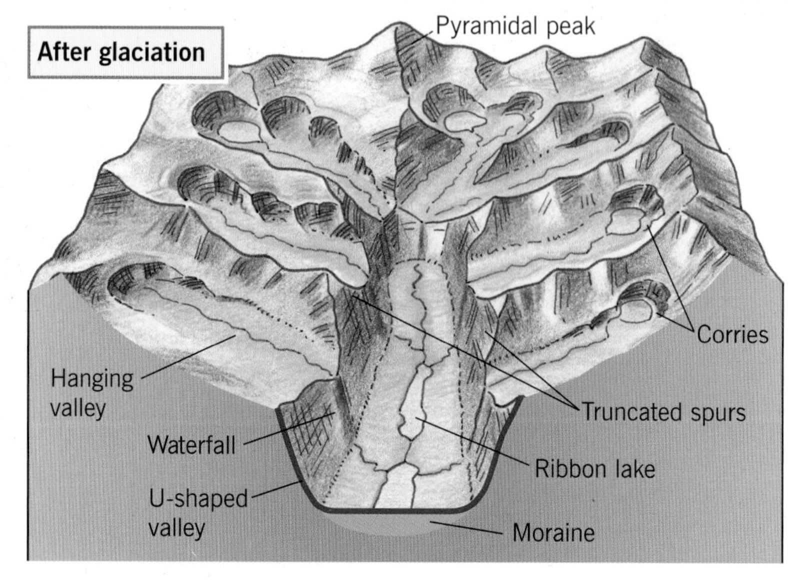

B Nant Ffrancon valley, Snowdonia, UK

C Glacial landforms

Truncated spurs are ridges that have been shortened or cut off by the moving glacier.

Hanging valleys are smaller valleys that have been eroded less than the main valley.

Ribbon lakes are long, narrow lakes that form in hollows in the main valley.

Erratics are boulders transported by ice and deposited often in an area of totally different rock.

Activities

1 What is a glacial trough and how does it form?

2 Drawing **D** shows a typical highland area that has been glaciated. The numbers refer to eight glacial landforms.

Match up the numbers with the correct landform. Choose from the list below.

- arêtes
- ribbon lake
- glacial trough
- hanging valleys
- pyramidal peaks
- truncated spurs
- corries
- erratics

3 How does glaciation:

a turn a V-shaped valley into a U-shape

b turn interlocking spurs into truncated spurs

c result in the formation of many waterfalls?

D

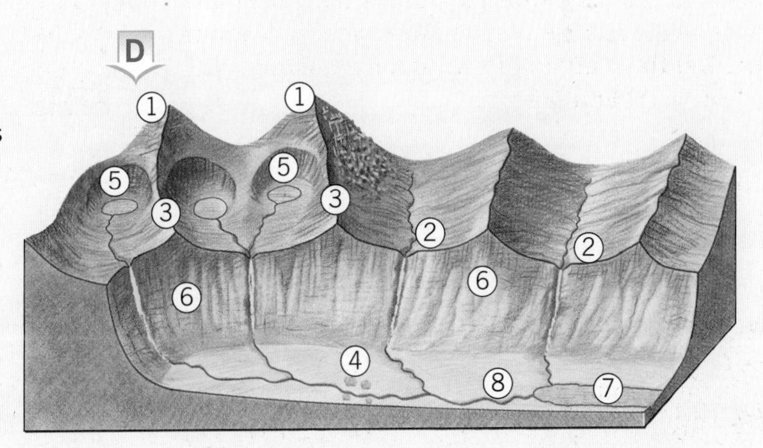

Summary

Ice can shape the land in many ways. Some landforms are shaped by glacial erosion while others are a result of glacial deposition.

2 Economic activity

What is economic activity?

What is this unit about?

This unit looks at the various types of work people do and describes the factors that affect the location of different industries.

In this unit you will learn about:

- the differences between primary, secondary and tertiary employment
- the distribution pattern of the main types of farming in Britain
- choosing the best site for a manufacturing industry
- the rapid growth of the tourist industry and the most visited places
- the growth and location of high-tech industries
- how employment structures change over time and between places.

Why is learning about economic activity important?

Work is an important part of our lives. It provides things that we need and by creating jobs it gives us the opportunity to achieve a higher standard of living and a better quality of life.

Learning about economic activity can help you:

- choose the sort of job that is most suited to you
- appreciate problems in providing our food and in making things that we need
- appreciate why industries have to be located in certain places
- choose places to visit on holiday and in your spare time.

A Rice cultivation in India

Which of the workplaces shown in photos **A**, **B** and **C**:

- might you expect to employ
  - most people
  - fewest people
- would you
  - most like to work in
  - least like to work in

Give reasons for your answer.

High-tech industry **B**

C Waikiki Beach, Hawaii

What types of economic activity are there?

Most people have to **work** to provide the things they need in life. Another word for the work they do is **industry**. There are many different types of work and industry. Together they are called **economic activities**. **Economic** means money and wealth.

The work people do can be divided into three main types. These are **primary**, **secondary** and **tertiary**. They are explained below.

- **Primary industries** employ people to collect or produce **natural resources** from the land or sea.
- Farming, fishing, forestry and mining are examples of primary industries.

- **Secondary industries** employ people to make things. They are usually made from raw materials or involve assembling several parts into a finished product.
- Examples are steel making, house construction and car assembly. **Manufacturing** is another name for this type of industry.

- **Tertiary industries** provide a service for people. They give help to others. No goods are made in this type of industry.
- Teachers, nurses, shop assistants and entertainers are examples of people in tertiary industries. This is sometimes called a **service** industry.

The proportion of people working in **primary**, **secondary** and **tertiary** activities is called the **employment structure**. The graphs in **D** show that the employment structure changes over periods of time.

In the UK before 1800 most people worked on the land while others often made products needed by farmers, such as ploughs, or produced by the farmer, such as bread. This was still part of the agricultural revolution.

During the 19th century people turned to industry, with an increasing number working in factories that produced steel, ships and machinery. This was the time of the industrial revolution.

Further changes occurred during the 20th century and up to the present day. Farming and industry have become highly mechanised and need fewer workers. Most people now work in service activities such as in schools, hospitals, shops and transport.

Since the 1980s a fourth group, **quaternary activities**, has been added to the employment structure (photo **E**).

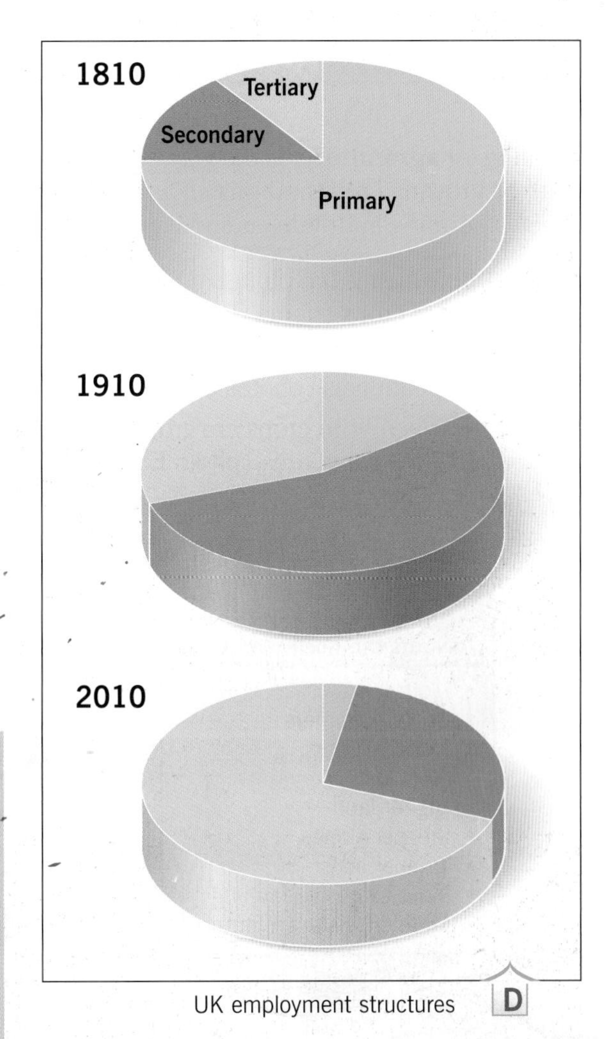

UK employment structures **D**

E

- **Quaternary industries** provide a high-tech service that carries out research and provides information and advice.
- Examples are financial advisers, research scientists and people working in micro-electronics.

Activities

1 Match the following beginnings with the correct endings:

Beginnings	Endings
Primary activities	make things from natural resources.
Secondary activities	collect natural resources from land or sea.
Tertiary activities	are mainly connected with information.
Quaternary activities	provide a service for people.

2 Look at the jobs listed in **F**. Sort them into primary, secondary and tertiary activities.

3 a Conduct a survey. Each person in your class should find out the job of one member of their family.
  b Sort the jobs into primary, secondary, tertiary and, if you can, quaternary.
  c Draw either a bar graph or a pie graph to show your findings.
  d Describe what your graph shows.

F

Job Shop

- TV presenter
- coal miner
- nurse
- shoemaker
- footballer
- forestry worker
- fire fighter
- bus driver
- carpenter
- police officer
- baker
- shop assistant
- fisherman
- quarry worker
- farmer
- pop singer
- builder
- sewing machinist

Summary

The three main types of industry are primary, secondary and tertiary. Recently, a fourth group, quaternary, has been added. The proportion of people employed in these groups changes over time.

What are the main types of farming in Britain?

Farming, or **agriculture**, is the way that people produce food by growing crops and raising animals. The main types of farming in Britain, as shown in map **A**, are:

- **arable**, which is the ploughing of the land and the growing of crops (photo **B**)
- **pastoral**, where the land is left under grass for the grazing of animals (photos **C** and **D**)
- **mixed**, when both crops are grown and animals are reared in the same area (photo **E**).

Farming, especially in Britain, is big business. Farmers must choose carefully the type of farming that is best for the place where they farm. Deciding which is best depends upon several physical and human factors.

- **Physical factors** are climate, soils and relief.
- **Human factors** include farm size, technology, machinery, distance from markets, market demand and transport.

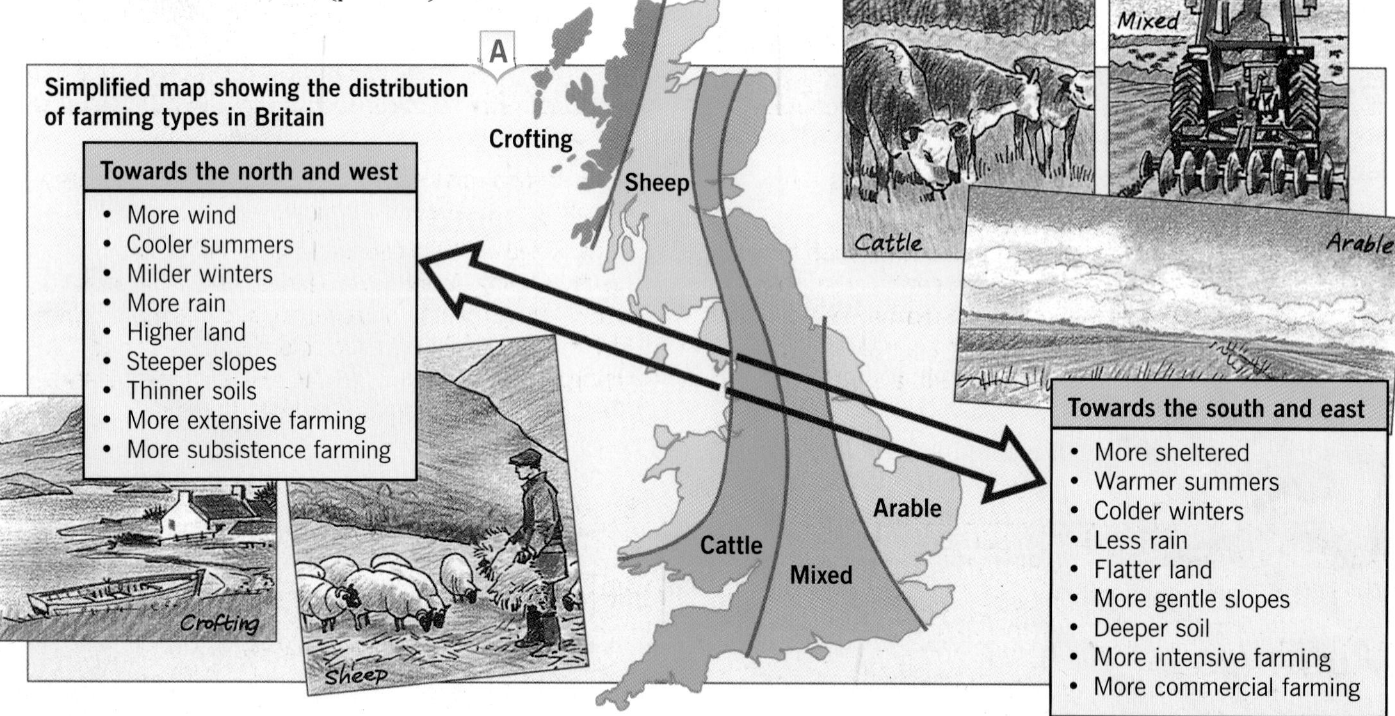

- Any climate
- Flat or gently sloping land
- Produces wool, lamb and mutton
- Fertile soil
- Warm, wet climate
- Raises animals and grows crops
- Produces milk or beef
- Hilly land
- Grows crops
- Warm, moist climate
- Flat land
- Little machinery
- Modern machinery

Activities

1 Describe each of the following farming types:

- arable
- pastoral
- mixed

2 Make a larger copy of table **F**. Put the statements from **G** into the correct columns. Some may be used more than once.

F

Farming type	Description	Needs
Arable		
Cattle		
Hill Sheep		
Mixed		

Summary

The main types of farming in Britain are arable, pastoral and mixed. The type best suited to an area depends upon several different physical and human factors.

B Arable farming

- Arable farms grow crops. These may be cereals like wheat, or vegetables such as potatoes.
- Arable farms need large areas of flat land, a deep fertile soil, warm and sunny summers to ripen the crop, some (but not too much) rain, and modern machinery for ploughing and harvesting. Crops that are heavy to move (potatoes) or soon go bad (strawberries) are best grown near to their market.

C Cattle farming

- Cattle farms raise either dairy cows for milk and dairy produce, or beef cows for meat.
- Cows need well-drained land that is either flat or gently sloping. Dairy cows need good-quality grass which grows best in a warm, moist climate and where it is neither too hot nor too cold. Good roads and refrigerated lorries help get the milk to market quickly. Beef cattle are reared further from markets.

D Hill sheep farming

- Hill sheep farms produce wool, lamb and mutton.
- Sheep are hardy animals and can graze on land that is too steep for other types of farming. They can also feed on poor-quality grass which grows on thin, infertile soils, and they can survive in any type of climate in Britain, including heavy rain and snow. Hill farms use little machinery and much of their produce goes to large urban areas.

E Mixed farming

- Mixed farms grow crops and raise animals.
- They need good-quality soil and gently sloping land for their cattle, crops and modern machinery. The climate should not be extreme – not too wet, too dry, too hot or too cold. Good roads for access to markets are also important.

What is a hill sheep farm like?

Beckside Farm (photo **A**) in the English Lake District is a typical hill sheep farm. Most of the farm is made up of high fell and steep and rocky hillside. The soils here are poor and will only support rough grasses and heather. The farm has only a small area of flatter, low-lying land. This has deeper soils but is often wet and difficult to cultivate.

Modern machinery is little used at Beckside Farm. The valley sides are too steep and inaccessible for tractors and machinery, whilst the valley floor, with its small fields and often waterlogged land, restricts the use of heavy machinery.

The climate in the area can be difficult for farming. Rainfall is heavy and low cloud and mist are common. Summers are cool and winters can be cold and windy. Snow may lie on the higher ground for several weeks (see photo **D** on page 31). Even sunshine amounts are lower than in the rest of the country.

The sheep on Beckside Farm spend most of their time on the high fells (diagram **B**). Life here is hard and only strong, healthy sheep survive. In winter, when conditions on the mountainside are uncertain, weaker sheep are brought down to lower pastures.

A

The small area of lowland is well used. The fields are ploughed each year and fertiliser added to improve the soil's fertility and structure. Oats, barley and turnips are grown during the summer and stored in the large barns. These are then given to the sheep as winter feed.

Beckside Farm earns most of its money in the autumn when lambs and four-year-old sheep are taken to the nearby market at Penrith. Wool is sold in summer but it is no longer a main source of income.

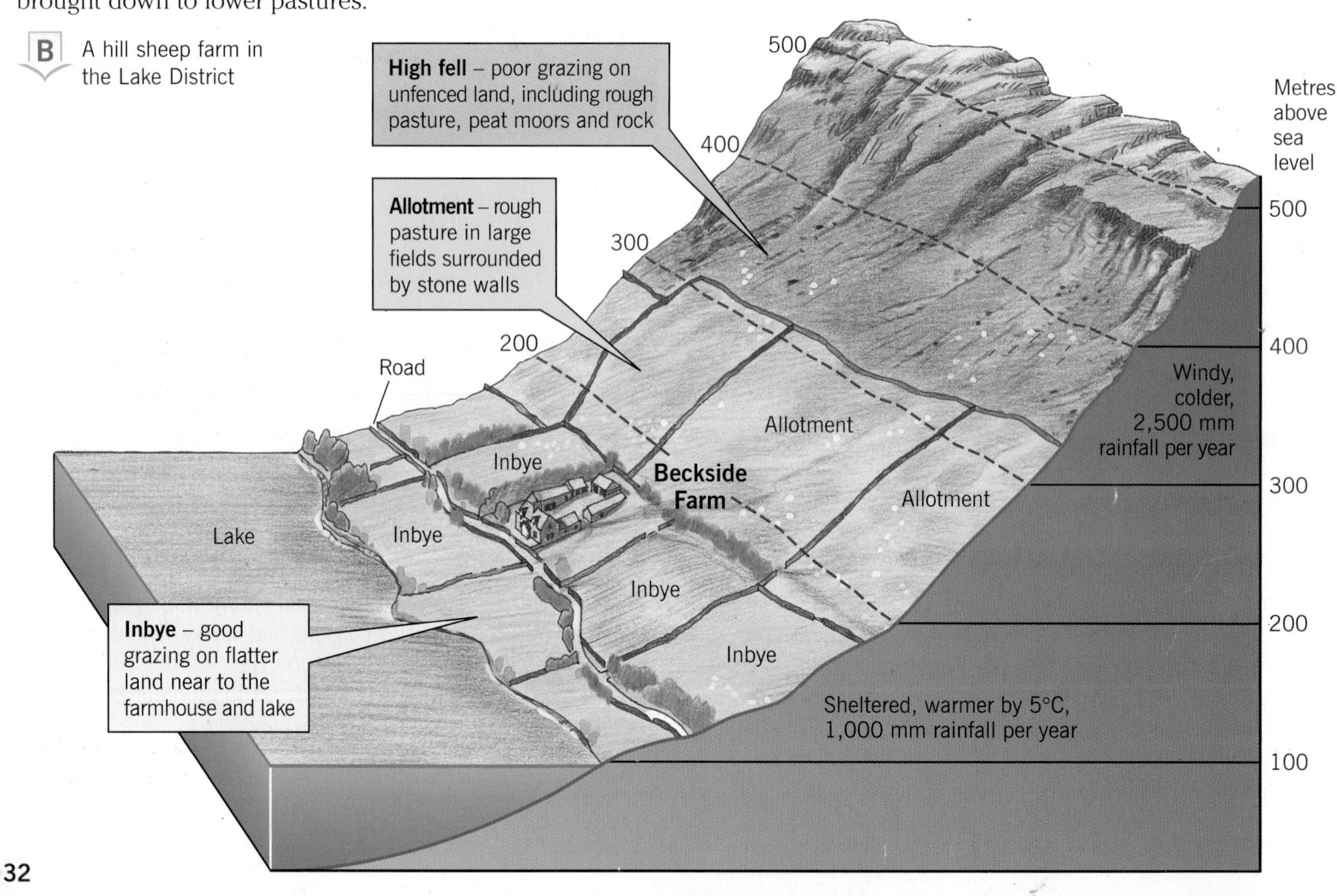

B A hill sheep farm in the Lake District

What is an arable farm like?

Photo **B** on page 31 and photo **C** on this page show Hawthorn Farm, a typical arable farm in East Anglia. The land here is very flat and slopes gently towards a river. Most of the fields have deep and fertile soils. These are easy to work and are well drained, although occasional flooding restricts the use of land near to the river.

East Anglia is one of the driest parts of Britain. Much of the rain falls in the growing season when it is most needed. Summers are generally warm with plenty of sunshine to ripen the crops. The cold winters have hard frosts which help to kill diseases and break up the soil, which assists ploughing.

The main problems for farmers in the area are occasional high winds and summer thunderstorms that can ruin the crops just before harvest time. Farmers cope with unusually dry summers by irrigating the land through a system of ditches, pipes and pumping units. Pumps are also used to move water from the fields and into the river when flooding occurs. Flooding is a problem in the area because the land is so flat and low-lying. Many farmers are concerned that floods happen more often now than they did in the past.

C

Much use is made of machines and chemicals. Hawthorn Farm has five tractors, two combine harvesters, muck spreaders, sprayers, ploughs, seed drills and a grain drier. There are six full-time labourers, and several casual workers are employed at different times of the year.

Hawthorn Farm is large and efficient, and uses modern methods. The farm is owned by a company and run by a professional manager. The company has the money needed to invest in new farming methods. Farming like this is called **agribusiness**.

Activities

1 Use map **D** to describe where Beckside Farm and Hawthorn Farm are located.

2 Make a fact file to show the main features of:
   a an arable farm
   b a hill sheep farm.

   Use the headings shown in table **E**.

D

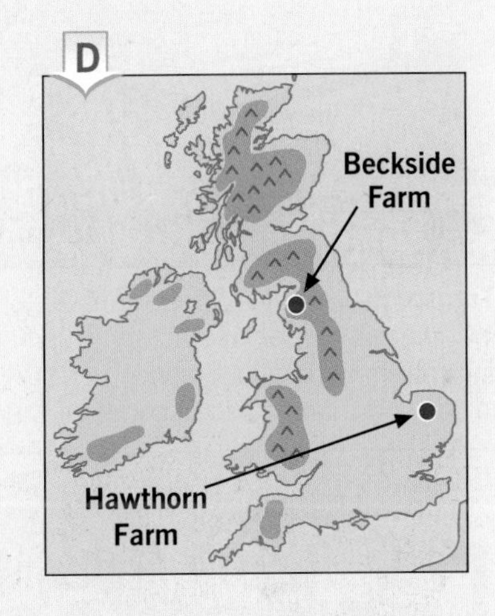

E

	Hill sheep farm	Arable farm
Location		
Relief		
Soils		
Temperature		
Precipitation		
Machinery		
Transport		
Markets		
Difficulties		
Source of income		

Summary

The difference in farming between a hill sheep farm and an arable farm is due to climate, soils, relief, transport and nearness to markets.

What is the best site for a factory?

Before building a factory a manufacturer should try to work out the best site for its location. It is unusual to find a perfect site for a factory. Indeed, if there was a perfect site someone else would probably already be using it. Deciding on the best available site depends on several things. Six of these are given in diagram **A**.

Choosing the right site – the iron and steel industry

Industries near to raw materials

Britain was the first country in the world to become **industrialised**. Industrialisation began about 200 years ago after the discovery that coal could be used to produce steam and that steam could be used to work machines. Machines did many of the jobs previously done by people.

In those days transport was poor. There were no lorries or trains, no motorways or railways. Coal and other raw materials were heavy and expensive to move. This meant that most early industries grew up on Britain's coalfields. The most important industry became the production of iron and, after 1856, steel.

The iron and steel industry

In the early 19th century large amounts of coke (coal) were needed to smelt (melt) iron ore. These, together with limestone, were found nearby in the valleys of South Wales. Water was also important. It provided both power from waterwheels and transport by canal.

Iron factories were built on the flat but narrow valley floors found in South Wales. They also had to be near to urban areas which provided the large number of unskilled workers that were needed. When steel was produced after the 1850s, it was sent to many parts of Britain to be made into ships, trains, bridges and textile machinery.

B Port Talbot – an ideal site for a steelworks?

Today Port Talbot is the largest of the few remaining steelworks in Britain. It produces mainly sheet steel for the car industry in England. As local coal and iron ore have been used up, they have to be imported from overseas. As photo **B** shows, Port Talbot has its own docks (top left) and is built on a large area of flat land close to the M4 motorway (bottom right) and mainline railway. Power comes from the national electricity grid and the steelworkers, now skilled, from nearby towns.

Activities

1 Make a list of the six terms shown in bold on diagram **A**. Match each term with the correct endings given in the list below.

- ... is needed to move raw materials, workers and goods.
- ... is a place where manufactured goods are sold.
- ... is where the factory is built.
- ... is needed to work the machines.
- ... are natural resources from which goods are made.
- ... is people who work in the factories.

2 Using photo **B** and the questions in diagram **C**, explain why Port Talbot is a good location for a steelworks.

3 Choose a factory near to where you live. Using diagram **C** as a guide, explain why it grew at that site.

C

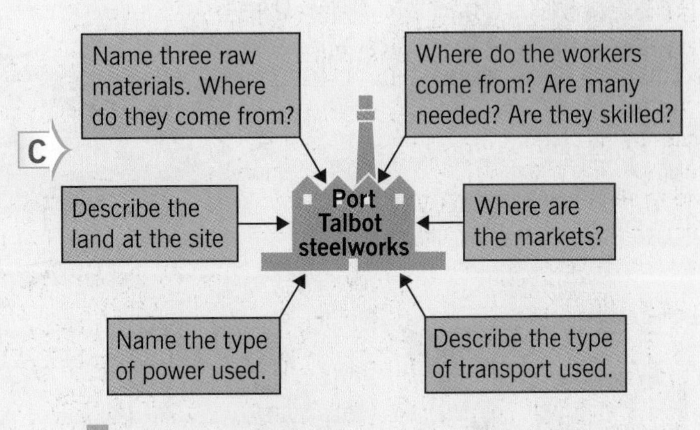

Summary

In choosing the best site for a factory, a manufacturer should consider transport and the nearness to raw materials, power sources, workers and markets for the goods produced.

Choosing the right site – the car industry

Industries near to markets

As raw materials are used up, and as transport improves, then modern factories tend to locate in areas where many people live. This is mainly because present-day industries need large markets in which to sell their goods. The car industry is an example of an industry that builds new factories near to markets.

The car industry

A modern car consists of many small parts. Each part is made in its own factory. If the factories making these parts are all close together then it is easier and cheaper for the car manufacturer to **assemble** (put together) all of these parts. If large towns are nearby then workers from these towns can make and assemble the parts and, hopefully, buy many of the finished cars. Transport is important for moving car parts, assembled cars and workers. Map **A** shows where most people in Britain live and where the largest car assembly plants were located in 2013. Over time some car factories have closed while others have opened. Most have had a change in ownership in recent years.

Today industrial growth is more likely in those areas where there are most people. In these places new factories are opening, jobs are easier to get, and more care is taken of the environment.

A Location of car assembly plants in 2013

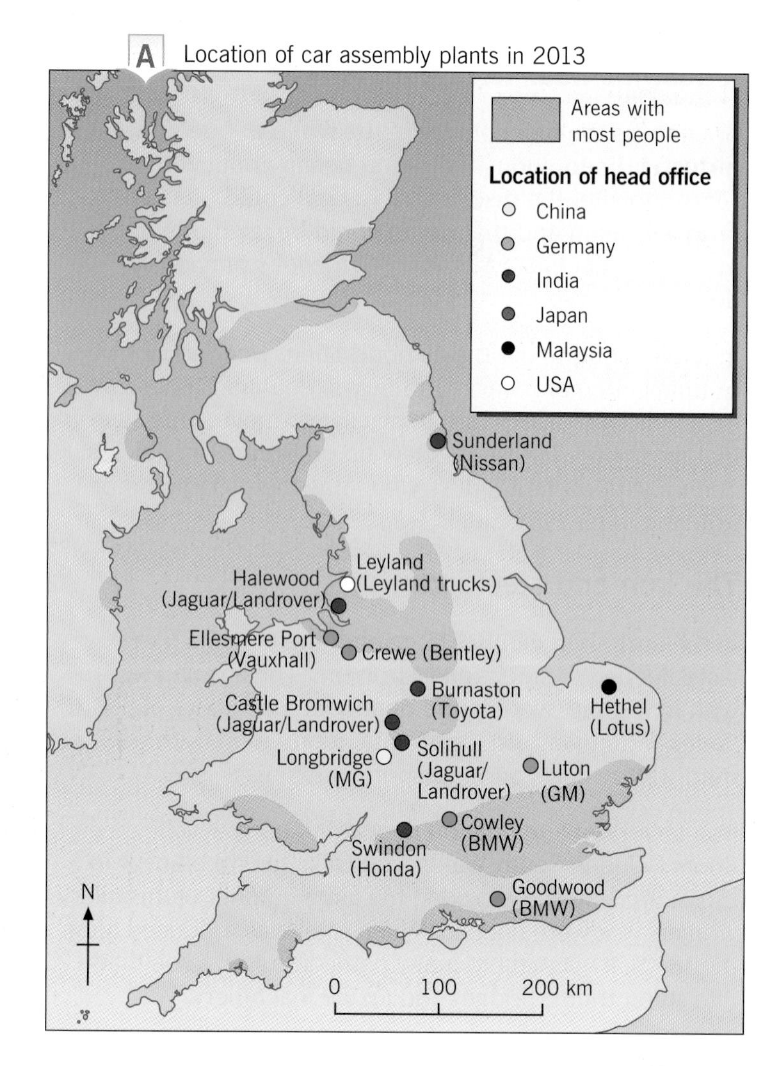

B A car assembly line

Toyota is Japan's largest car company. In the early 1990s it decided to build a new car manufacturing plant at Burnaston near Derby. The plant opened in 1992 and in 2012 was producing 130,000 cars a year with a workforce of 3,800 people. Most Toyota cars are sold in the UK and Europe. Some are even transported to Japan to be sold there.

Toyota uses a **just-in-time** system of manufacture where components (car parts) are supplied to the assembly line just minutes before they are needed. Expensive parts do not have to be stored on site so costs are reduced. Just-in-time needs a good transport system for it to work. Map **C** shows some reasons why Toyota chose Burnaston.

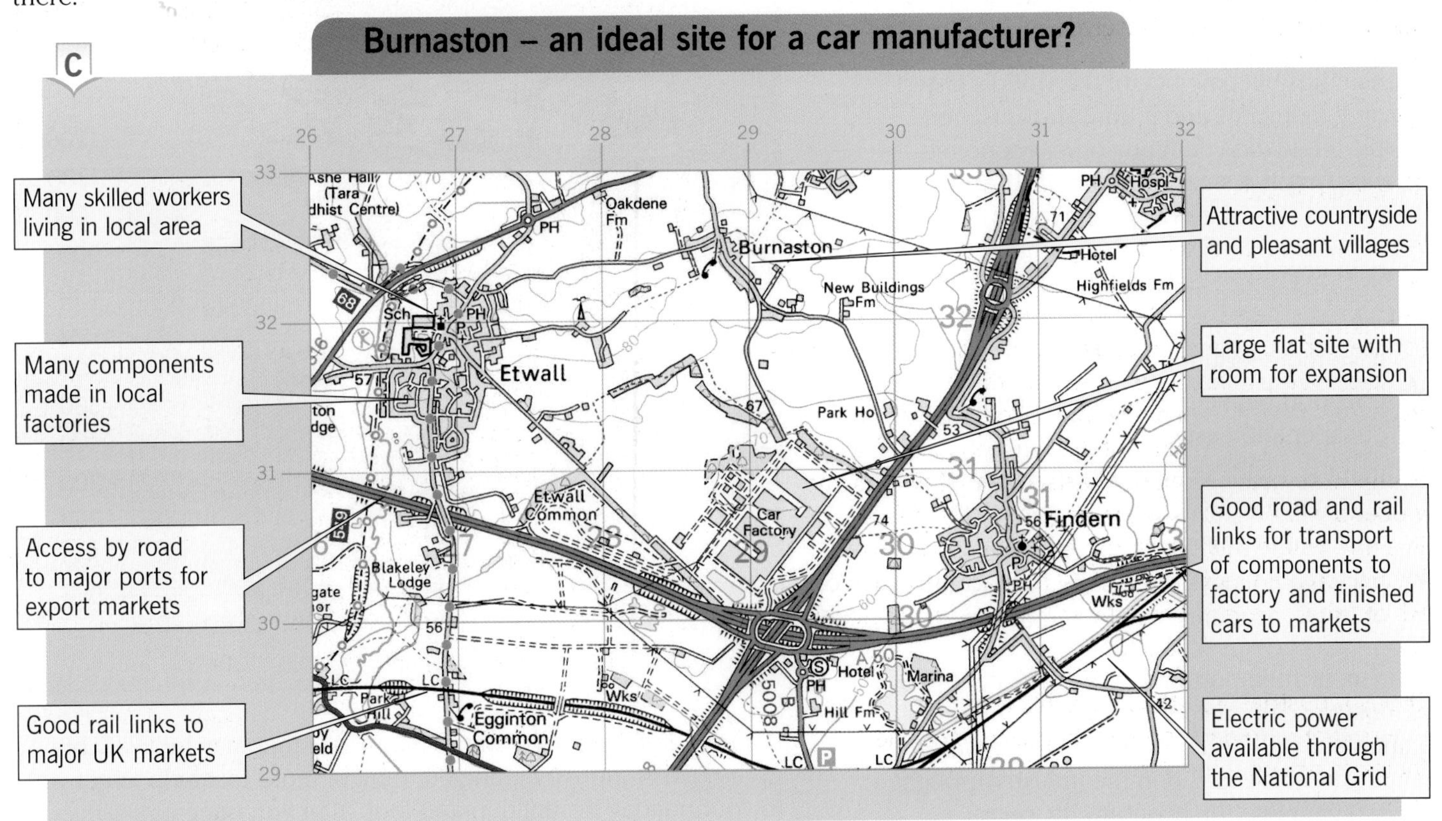

Activities

1 Of the six statements shown in drawing **D**, four are correct. Write out the correct ones.

2 Car factories are usually located close to large towns. Give at least two reasons for this.

3 What are the advantages and disadvantages of the just-in-time system?

4 Explain why Burnaston is a good site for a car factory. Use the headings in fact file **E**.

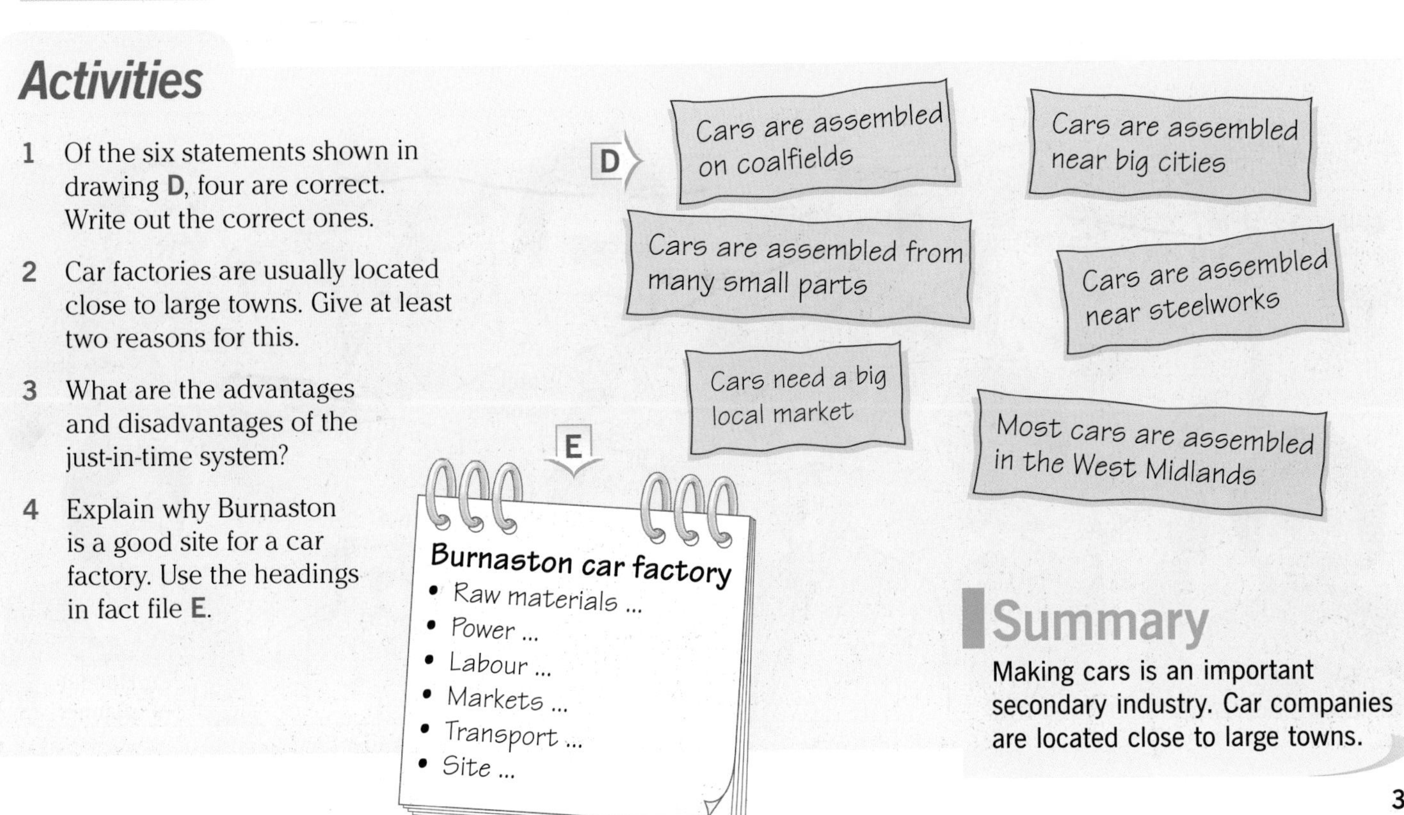

Summary

Making cars is an important secondary industry. Car companies are located close to large towns.

What is the tourist industry?

Tourists are people who travel for pleasure. The **tourist industry** helps these people by getting them to the places they want to visit, and then looks after them while they are there to ensure they relax and enjoy themselves.

Tourism has become one of the world's fastest-growing industries and it now employs more people worldwide than any other economic activity. Graph **A** shows this increase, and clearly there is no sign of it slowing down.

Tourism continues to grow because:

- many people are earning more money and have more leisure time
- transport between places has become faster, cheaper and easier
- there is a wider range of holiday destinations and activities which people may have seen on TV or in adverts in the media
- package holidays often include the total cost of travel, accommodation and meals.

Tourism is an important factor in the economy of most developed countries. In Britain, for example, in 2012 one person in twelve was employed in tourism, and the industry brought £35.9 billion into the economy. In many developing countries it is often seen as the best way to obtain income and create jobs.

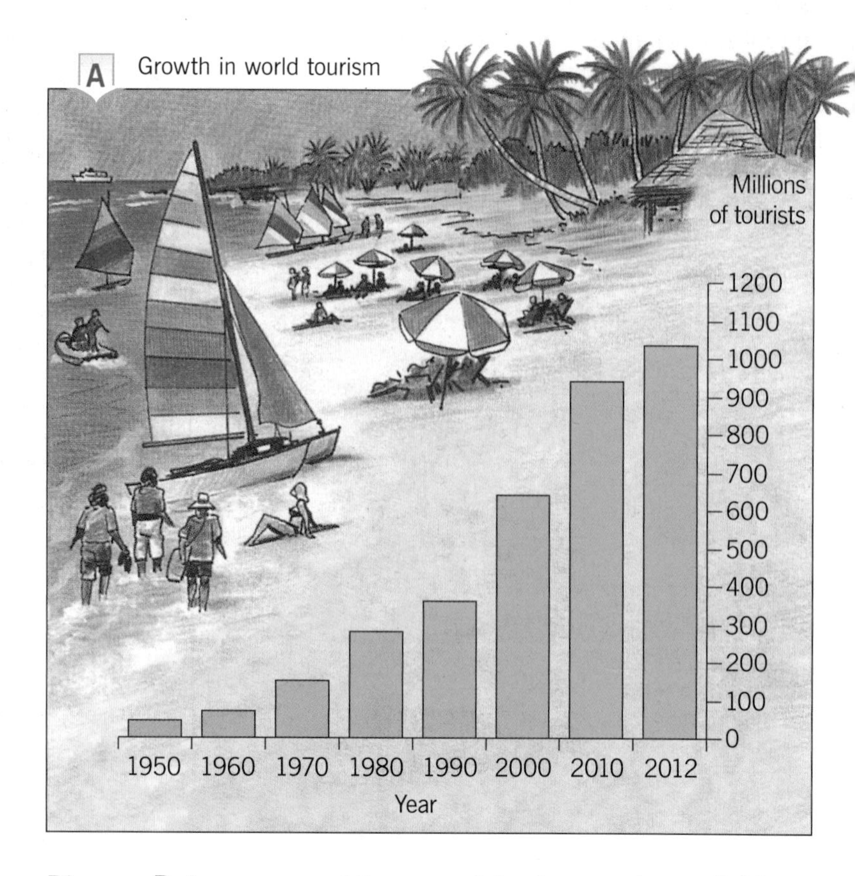

Diagram **B** shows some of the many jobs that may be available in the tourist industry. Can you think of more that help to get you to your holiday destination and look after you once you arrive there?

Tourism can bring advantages to places that attract many visitors. To them it becomes an important source of money and jobs. Places that were once quite poor, even in Mediterranean Europe, now have a higher standard of living and quality of life for their local people.

C Advantages and problems of tourism

Advantages

Can bring improvement in:
- **standard of living** – by creating jobs in hotels, restaurants, cafés, bars, discos, making and selling souvenirs, acting as tour guides and bus drivers, and providing entertainment and farm produce
- **quality of the environment** – by enabling new schools, hospitals and roads to be built.

Problems

Can affect the local **environment** and its scenery, vegetation and wildlife:
- **tourists** can cause overcrowding, noise and litter as well as getting drunk and using drugs
- **local people** find jobs are seasonal, unskilled and poorly paid, shopping gets more expensive and the local culture and traditions are lost.

At the same time tourism can create problems, including those that can harm the environment and affect the way of life of local people (diagram **C**).

Activities

1 Describe what graph **A** shows.

2 a Which of the jobs in diagram **B** would you:
- most like to do
- least like to do?

Give reasons.

b Make a list of at least 15 jobs that might be found in a large tourist holiday centre.

3 There are plans to build a large new hotel complex like the one in photo **D** on the coast at **E**, in an area where there are few jobs.

Give three reasons why some people are:
- in favour of the plan
- against the plan.

D

E

Summary

Tourism is one of the world's fastest-growing industries. It can bring wealth, jobs and amenities for local people but it can also create problems for them and for the environment.

Where do the tourists go?

Every year, more and more people are taking holidays, more places are visited and there are more things to do. Transport has improved. It is also both cheaper and quicker, making it possible to visit almost anywhere in the world.

Even so, countries in Europe are by far the most popular tourist destinations, especially Spain and Italy with their hot, dry Mediterranean summers. Many people also visit France, Germany and the UK to see their historical attractions. France and Italy are also popular for skiing in winter (table **A**).

In more recent years a group of developing countries lying within the tropics have benefited from tourism. These countries include Kenya, Sri Lanka, Thailand, Malaysia and some Caribbean islands.

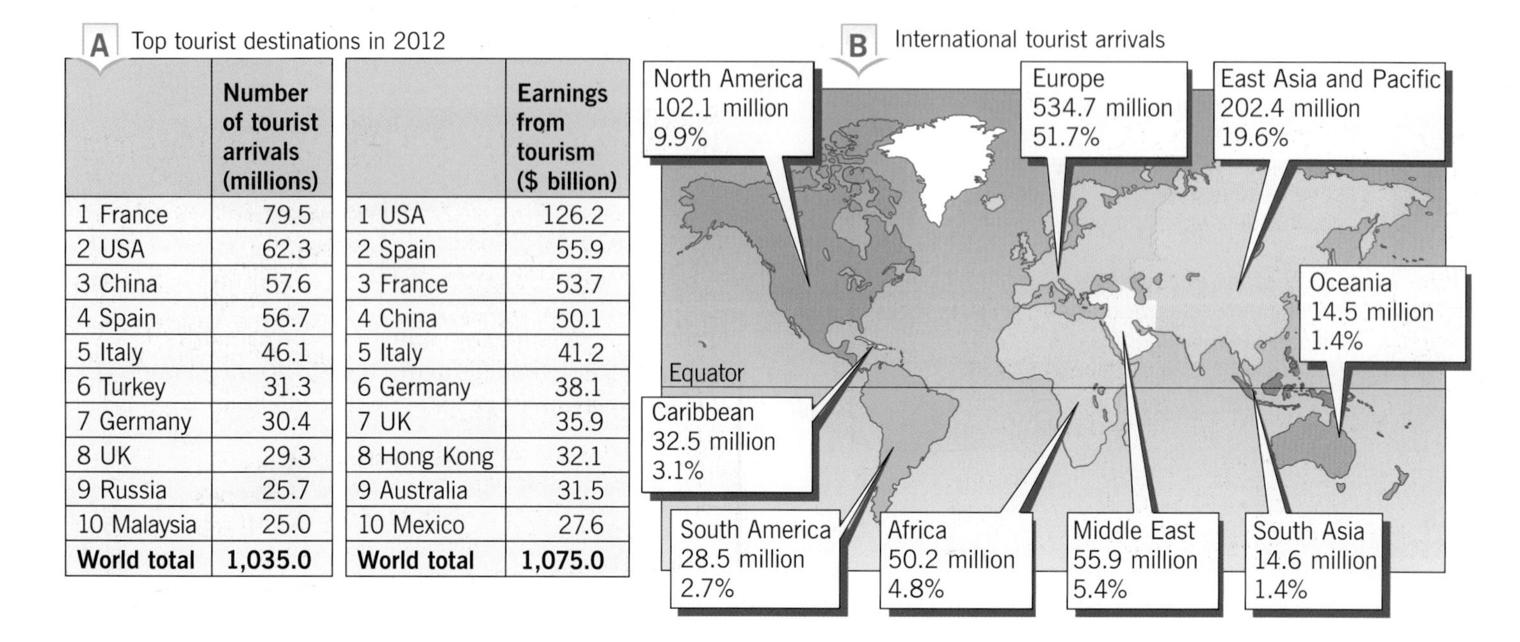

A Top tourist destinations in 2012

	Number of tourist arrivals (millions)		Earnings from tourism ($ billion)
1 France	79.5	1 USA	126.2
2 USA	62.3	2 Spain	55.9
3 China	57.6	3 France	53.7
4 Spain	56.7	4 China	50.1
5 Italy	46.1	5 Italy	41.2
6 Turkey	31.3	6 Germany	38.1
7 Germany	30.4	7 UK	35.9
8 UK	29.3	8 Hong Kong	32.1
9 Russia	25.7	9 Australia	31.5
10 Malaysia	25.0	10 Mexico	27.6
World total	**1,035.0**	**World total**	**1,075.0**

B International tourist arrivals

As well as more places being visited around the world, there are more holidays for tourists to choose from (diagrams **D** and **E**). People are also looking for more adventurous and active holidays, such as hot-air ballooning, whitewater rafting, polar bear watching and mountain trekking.

So how do we decide where to go? Most of us are influenced by adverts or television programmes, or learn about places from friends who have visited and liked certain places – but there is more to it than that. You must learn to ask questions and try to find our more about a place before making a choice. Photo **C** shows some questions to think about when choosing your holiday.

C

D

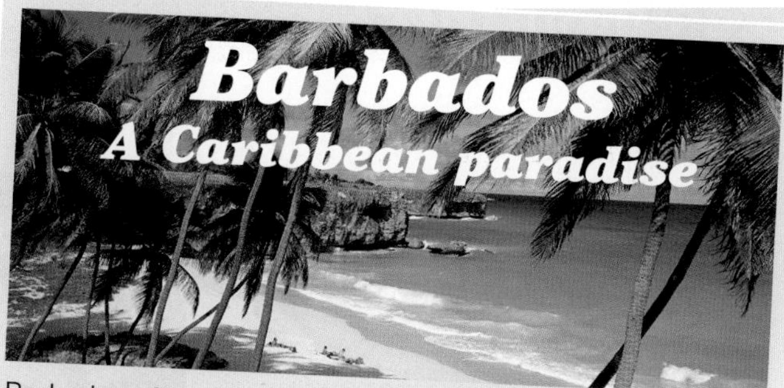

Barbados offers you hot sun and warm sea all year round. There are fine sandy beaches, swaying palm trees, lovely hotels and fine restaurants. Have a relaxing beach holiday or enjoy excellent watersports activities.

Prices from £1,890 for two weeks

Kenya has it all – great weather, spectacular scenery and exciting wildlife. Enjoy a safari and see lion, elephant, giraffe, zebra and many more. Then soak up the sun and laze on the unspoilt beaches of the Indian Ocean.

Florida

Glorious sunshine and hot all year. The wonderful world of Disney, the Epcot Centre, Universal Studios and Wet 'n' Wild. Fun and entertainment for all the family. Then chill out on the beach and maybe swim with the dolphins.

Activities

1 Look at map **B**.
   - **a** List the regions in order of popularity.
   - **b** Which region was by far the most visited?
   - **c** Suggest reasons for its popularity.

2 Twelve different types of holiday are listed in **E**. For each one, suggest where you would go for that type of holiday.

3 Where would you suggest as a holiday destination for:
   - **a** a family with three young children
   - **b** a retired couple?

   Give reasons for your answers.

4 Given a choice, where in the world would you like to go and what would you do there?

   Give reasons for your answer.

E

What type of holiday do we want?

- Famous buildings
- New cultures
- Beach resorts
- City breaks
- River/ocean cruises
- Wildlife
- Works of art
- Different foods
- Ski resorts
- Theme parks
- Spectacular scenery
- Better climate

Summary

Cheaper, quicker and easier travel has made it possible to travel almost anywhere in the world. Deciding where to go needs careful thought.

What are high-tech industries?

High-technology or **high-tech industries** make products such as microchips, computers, mobile phones, pharmaceuticals (drugs) and scientific equipment. They have been the growth industry of recent years and now provide more than 25 per cent of the UK's manufacturing jobs.

High-tech companies use the most advanced manufacturing methods. They put great emphasis on the research and development of new products and employ a highly skilled and inventive workforce. Most are huge organisations with offices and factories throughout the world. The UK electronics industry, for example, is controlled almost entirely by foreign companies, mainly from Japan and the USA.

Firms that make high-tech products often group together on pleasant, newly developed **science** or **business parks**. All firms on a science park are high-tech and have direct links with a university. Business parks do not have links with universities and may include superstores, hotels and leisure centres. There are many more business parks than there are science parks.

Both are located on edge-of-city **greenfield sites** although some business parks are found in inner-city areas that have been redeveloped, such as London Docklands. Photo **B** shows the sort of building to be found on a science park.

A Some high-tech companies

B Cambridge Science Park, UK

Diagram **C** shows some of the advantages to industry of science and business parks. Others include:

- They should be near to a main road on the edge of town for easy access.
- Nearby firms can exchange ideas and information.
- Leisure facilities and support services may be shared.
- A pool of highly skilled workers can be built up.

There can, however, be some disadvantages. For example:

- An over-use of cars can cause traffic congestion at busy times.
- Edge-of-town sites can be far from shops and services in the town centre.
- Firms may prefer to be by themselves so as to keep new ideas a secret.
- At times it may be difficult for firms to find enough skilled workers.

C Advantages of a business park

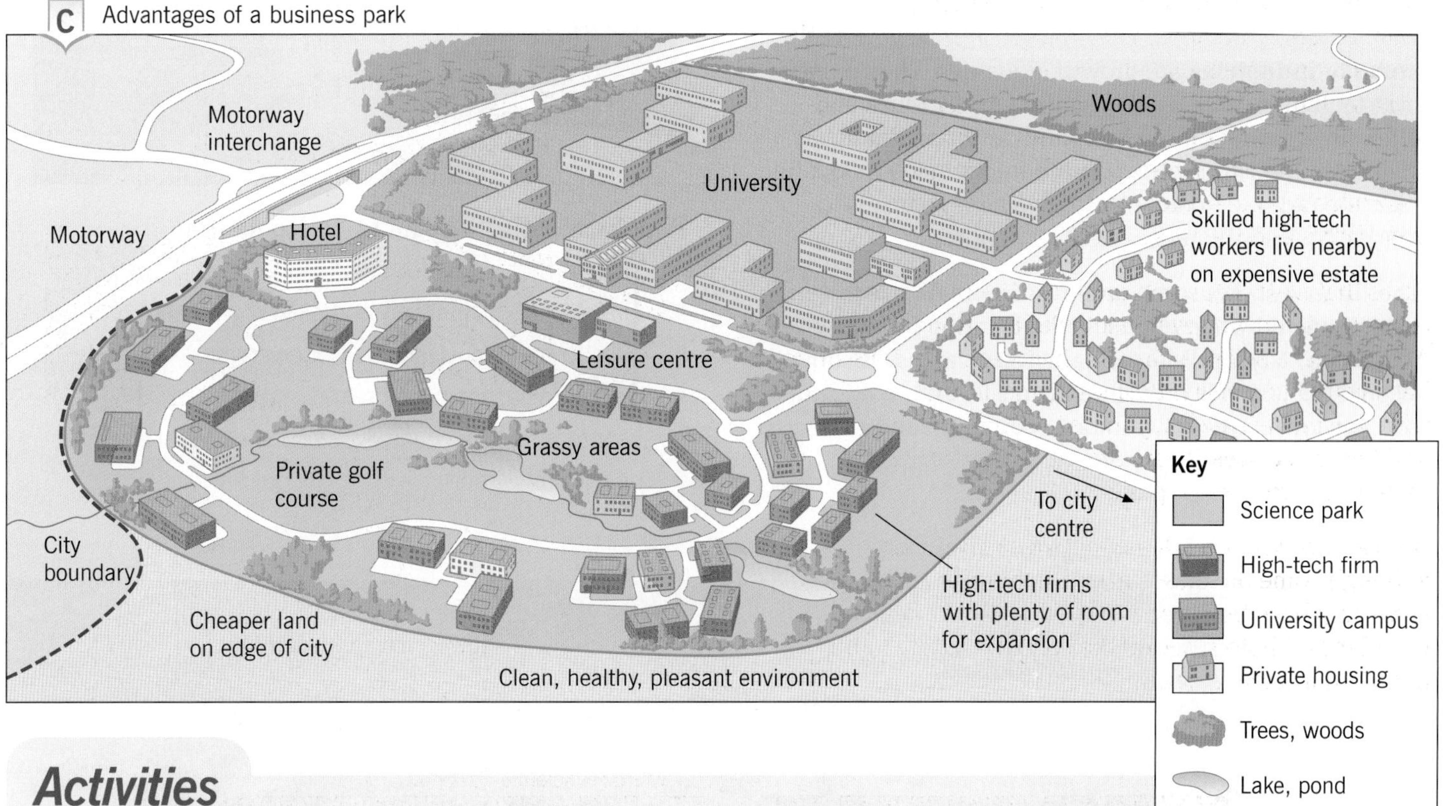

Activities

1. a What are high-tech industries?
   b Name five high-tech products that you have used in the last week.
2. Complete table **D** to show the differences between a science park and a business park. Choose your answers from the following pairs:
   a many/very few
   b university links/no university links
   c high-tech firms/high-tech firms, shops, hotels and leisure centres.

D

	Science park	Business park
a		
b		
c		

3. Using photo **B** and diagram **C**, give six reasons why a high-tech firm should locate on a science park. In your answer you should mention each of the following:

   transport, price of land, the environment, people's health, leisure facilities, exchanging ideas with people from other firms

4. Give two disadvantages which may arise from firms locating in the same place.

Summary

High-tech industries use advanced scientific techniques. They often locate on edge-of-town science or business parks.

Where are high-tech industries located?

Industries like shipbuilding, steelmaking, chemical manufacture and textiles used to be very important in the UK. They employed hundreds of thousands of people, mainly in the north and west of the country. These industries are now in decline and large numbers of jobs have been lost as companies have closed down or reduced their output. Industries in decline are often called **sunset industries**.

Sunrise industries are growth industries. They include high-tech industries which use modern factories and often have their own research and development units. High-tech industries have a much freer choice of location than the old traditional industries where nearness to market and raw materials were so important.

The list in **A** shows some of the factors that have to be considered when choosing the location for these industries. A typical site for a high-tech factory is shown in photo **B** which is on a greenfield site, on the edge of a country town, near to a motorway junction and close to a university.

A

High-tech industries like to be located:

- near to motorways or good roads
- near to highly qualified and skilled workers
- near to research facilities in universities
- near to pleasant housing and open space
- near to attractive countryside and good leisure facilities
- near to an airport for international links.

B

Although high-tech companies have factories and research centres in many parts of Britain, three main areas have become particularly important. These are:

1. The M4 corridor following the motorway westwards from London ('Silicon Strip')
2. 'Silicon Glen' in central Scotland
3. 'Silicon Fen' in and around Cambridge.

These locations are shown on maps **C** and **D**.

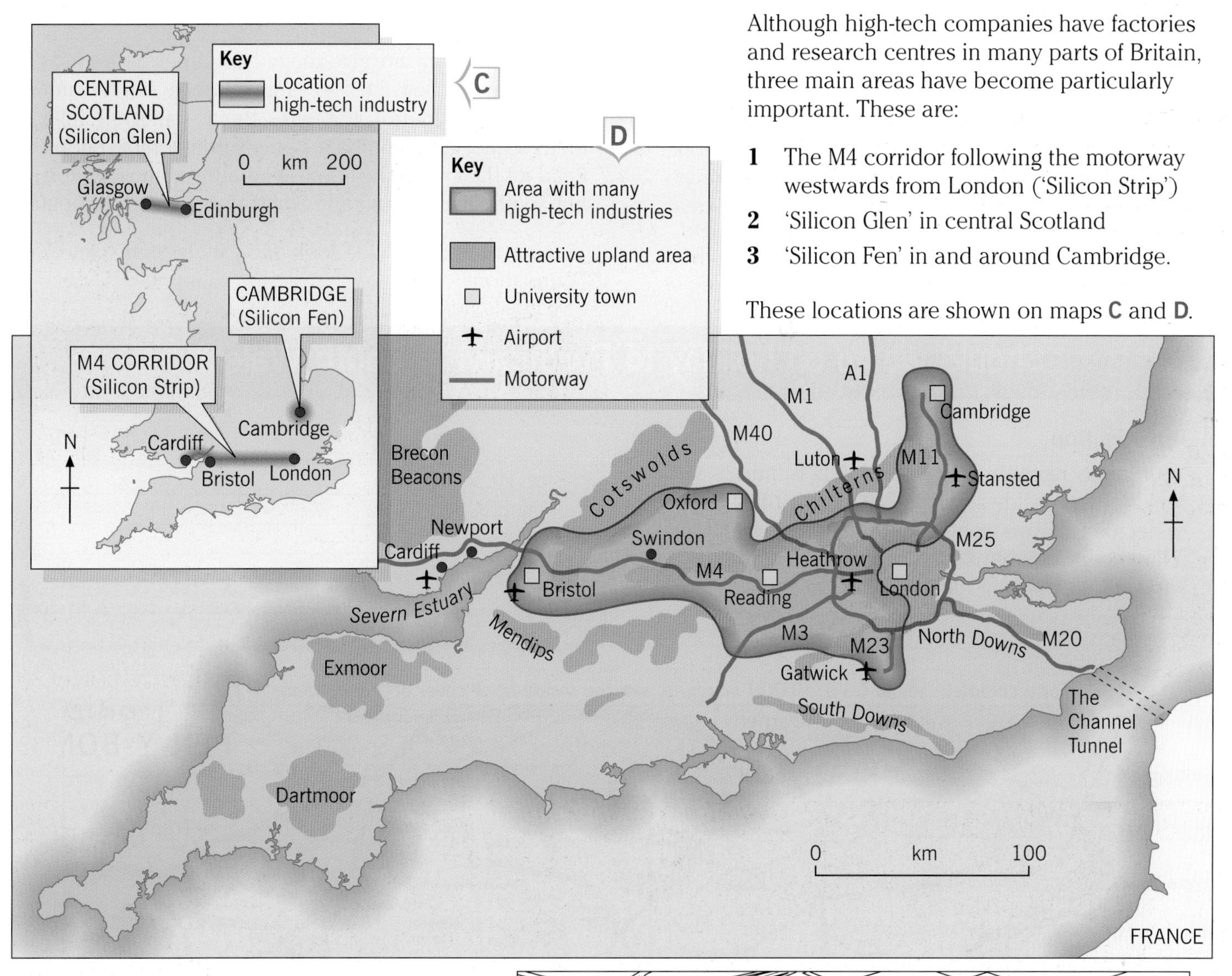

Activities

1. What are the differences between sunset and sunrise industries?
2. Drawing **E** is a sketch of photo **B**.
   - **a** Make a larger copy of the sketch.
   - **b** Colour lightly in pencil: the housing area red, the industrial area brown, the main roads yellow and the countryside green.
   - **c** Label the features that make the area a good site for a high-tech factory. Write about 10–20 words of explanation for each one.
3. Use the information on map **D** to explain why the M4 corridor is a good location for high-tech industries.

Summary

High-tech industries can locate in a wide variety of places. Most are found in central Scotland and the south and east of England.

The economic activity enquiry

Tourism is an example of a tertiary industry, which means people employed in it provide a service. Earlier in this unit, on page 38, you learnt that tourism is one of the world's fastest-growing industries and employs thousands of people worldwide. Tourism not only provides jobs, it also it brings much needed money to a country, a region or a local area.

Most British people now take at least one holiday a year. The main holiday for most people is likely to be outside of Britain, and many travel to the hot countries bordering the Mediterranean Sea. Even so, large numbers still remain in the United Kingdom. City breaks, theme parks and hilly areas attract large numbers, but in summer most people head for the coast. The South West attracts most visitors, but South East England and South Wales are also popular.

You may find it useful to look back at chapters 6 and 7 of Foundations.

Should we go on holiday to Porthcawl in South Wales?

1 Introduction

Your family have decided to take next summer's holiday at a seaside resort in Britain. Your first task is to decide why you want to go to the coast. First make a large copy of table **A**, and complete it by adding what attractions your ideal resort should have.

A

Physical attractions	1
	2
	3
Human attractions	1
	2
	3

For a complete key to OS 1:50,000 mapping, see the inside back cover of 'Foundations' **B**

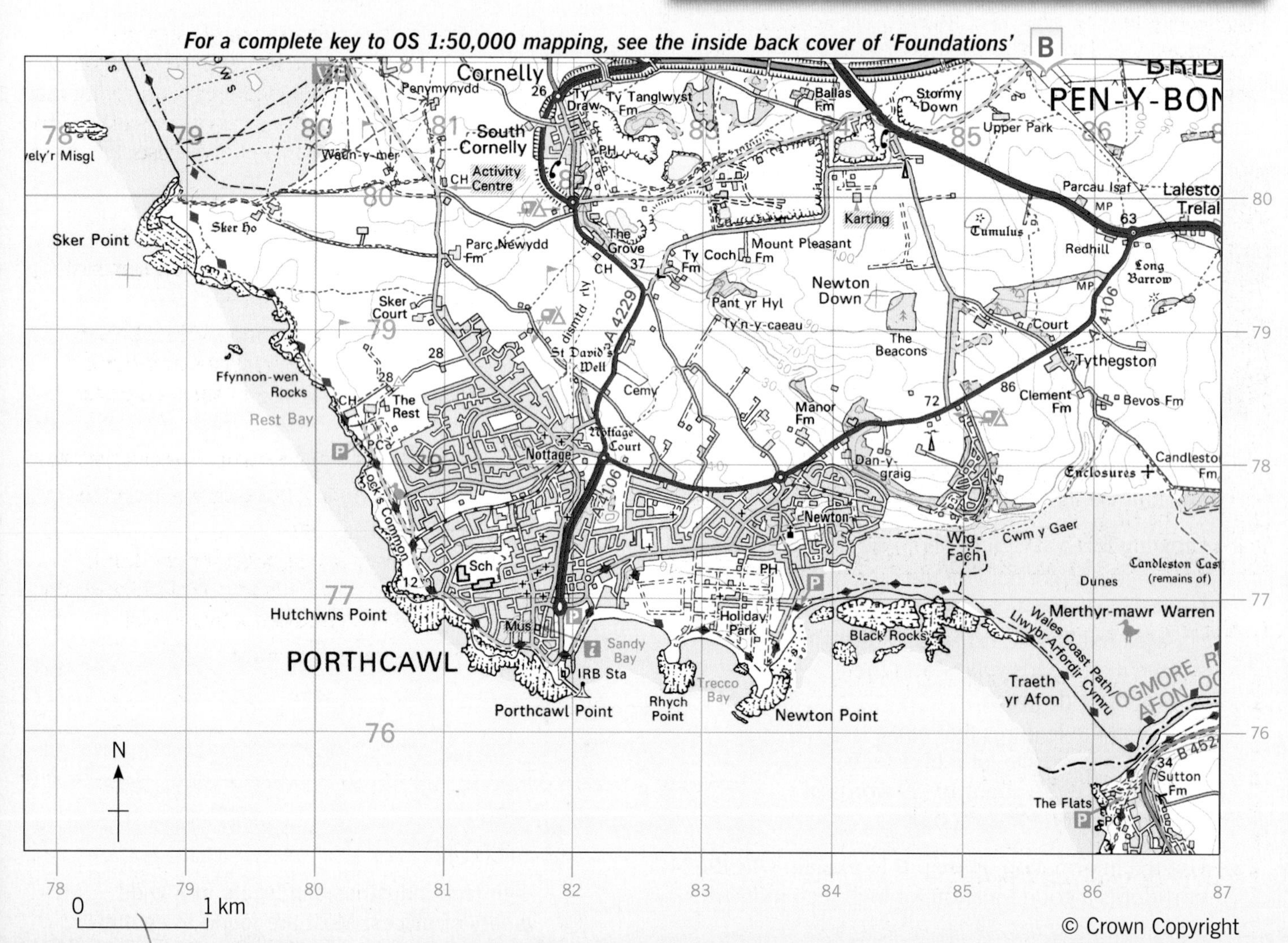

2 Map-reading skills and photo interpretation

You need to check that you can remember how to use your map-reading skills, and how to interpret an aerial photograph. You can do this by looking at the Ordnance Survey map **B** and the aerial photograph **D**. Then answer the questions in table **C** and beneath the aerial photograph.

C

Direction	**a** In which corner of the map is Sker Point? **b** In which direction is Newton Point from Sker Point?
Distance	What is the distance: **a** from east to west across the map **b** between the roundabouts at either end of the A4106 road?
Height	What is the highest: **a** numbered contour **b** spot height?
Four-figure grid references	Give the four-figure grid reference for: **a** the settlement at Newton **b** Sandy Bay.
Six-figure grid references	**a** What is at 837775 and at 820768? **b** What is the six-figure grid reference for the lighthouse at Porthcawl Point?

D

Look at the aerial photo above and answer these questions.

- **a** In which direction was the photo taken?
- **b** Name the bay at A.
- **c** Name the headland at B.
- **d** Name the headland at C.
- **e** Name the beach material at D.
- **f** Name the feature at E.
- **g** Give the road number at F.

3 Is Porthcawl the right place?

Porthcawl, a small coastal resort in South Wales, has been suggested as a suitable place to stay. You have been given an Ordnance Survey map of the area and an aerial photograph of the town. Like a detective, you should study the map and photo, looking for evidence that will help you decide whether Porthcawl is the ideal resort for you.

a Using map **B** and photo **D**, answer the following questions to help you decide whether Porthcawl fits your idea of an ideal seaside resort.

- What types of accommodation can you see?
- Which of your three ideal physical attractions are visible?
- Which of your three ideal human attractions are available?
- What physical attractions are there that you did not think of?
- What human attractions are there that you did not think of?

b Your family has a wide range of interests including walking, wildlife, sport, beach games and history. Using map evidence, and giving map references to support your answers, explain why Porthcawl would satisfy these interests.

4 Conclusion

a Draw a sketch map the same size as map **B**.

- Include the coastline and the approximate area covered by Porthcawl.
- In boxes around your map, add labels to show the physical attractions and human amenities suitable for tourists (see pages 108–111).
- Add arrows to point to their correct location.
- Colour the physical boxes in green and the human boxes in red.

b Write a summary saying whether you think Porthcawl is a seaside resort your family would like, might like, or would not like to visit.

Give reasons for your answer.

3 Population

People in the world

What is this unit about?

This unit is about people: where they live, how many there are and why they may move from one place to another.

In this unit you will learn about:

- where people live in the UK
- what affects where we live
- world population distribution
- how world population changes
- the causes and effects of migration.

Why is this population topic important?

Where we live and how many there are of us affects us all in some way or another. Learning about population can help us understand some of the problems facing our world. It will also enable you to develop your own views on how those problems may best be solved.

Learning about population can help you:

- understand why some places are more crowded than others
- choose where you would like to live
- understand the reasons for population growth and movement
- appreciate the problems resulting from population growth and movement.

A Dhaka, Bangladesh

B Dolomites, Italy

- Of the places shown in photos **A**, **B** and **C**, which:
  - looks the busiest
  - looks the poorest
  - would you most like to live in
  - would you least like to live in
  - would you like to know more about?

  Give reasons for your answers.
- Describe the likely feelings of the migrants shown in photo **D.**

C London, UK

D

Are we spread evenly?

There are about 61 million people in the United Kingdom but where do they all live? Map **A** shows this. It is a **population distribution** map and shows how people are spread out across the country. You can easily see that the population is not evenly spread out. There are some areas with a lot of people and some with very few. The south and the east seem to be most crowded, and the north and the west the least crowded.

The map uses **density** to show how crowded places are. Density is the number of people in an area. It is worked out by dividing the total population by the total area and is usually given as the number of people per square kilometre. Places that are crowded are said to be **densely populated** and to have a high population density. Places with few people are said to be **sparsely populated** and to have a low population density.

The most crowded places of all are towns and cities. In Britain today almost 9 out of 10 people live in a town or city. Some towns and cities are shown on map A and in table C. London is by far the largest and most densely populated city in the United Kingdom. More than 8 million people live there, and in the most crowded inner city areas there are up to 10,000 people in a square kilometre.

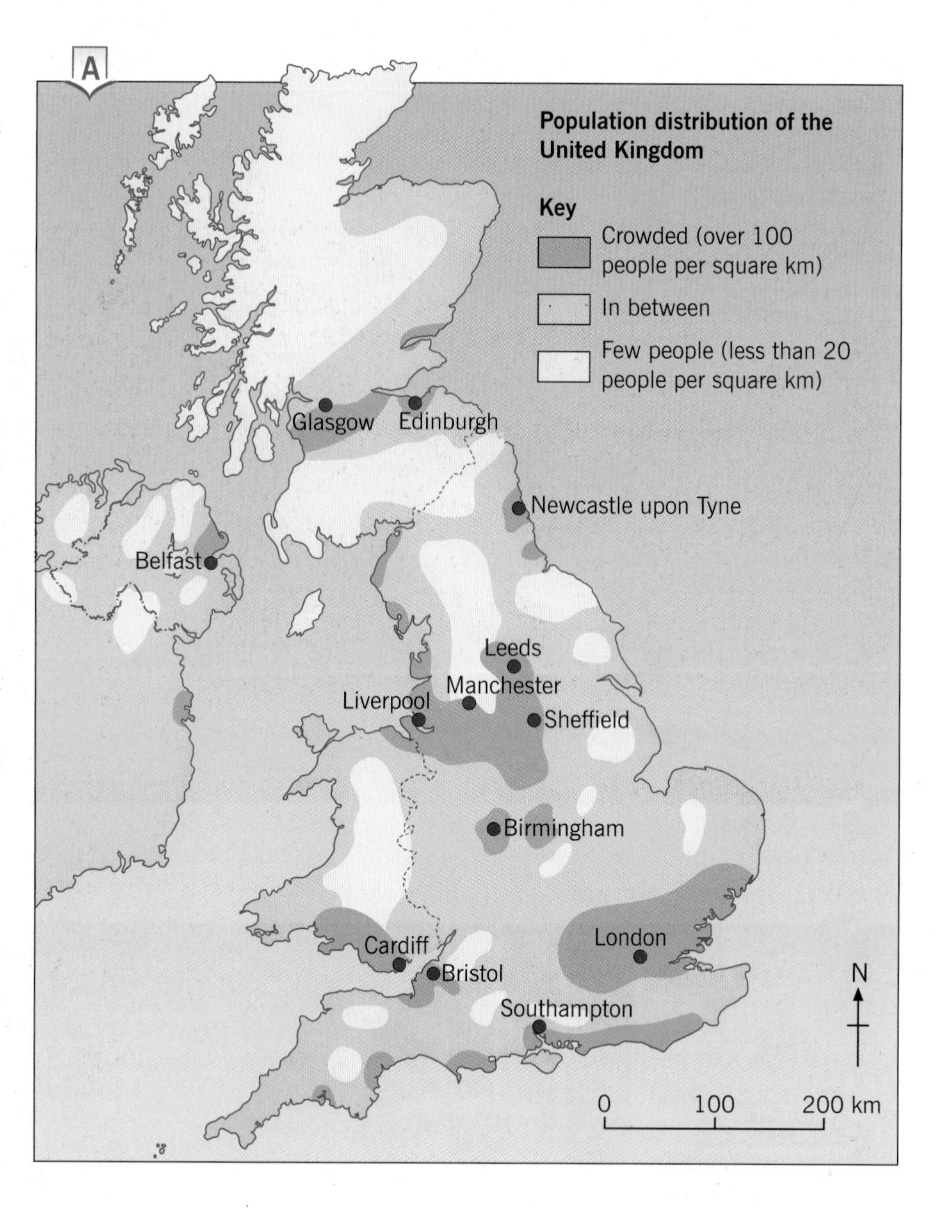

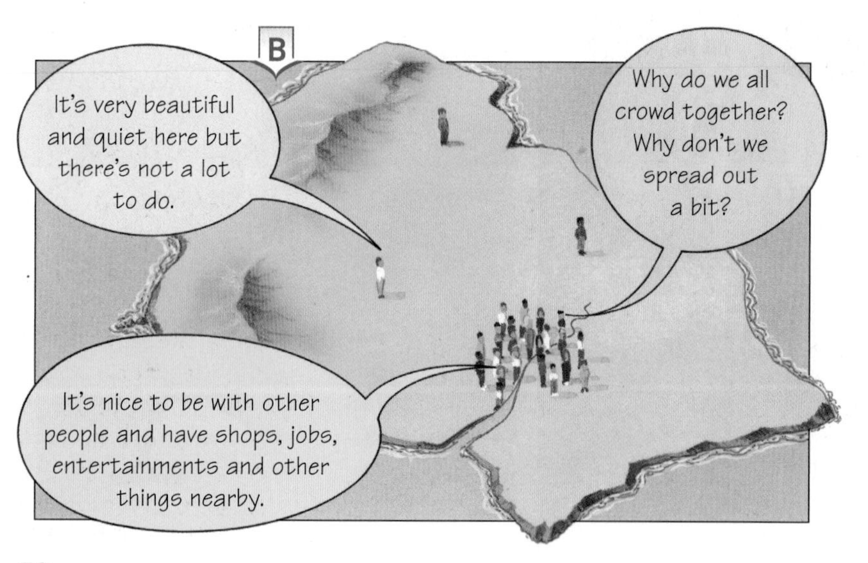

C

Population of some cities in Britain (figures are in thousands)

City	Population	City	Population
Belfast	288	Leeds	772
Birmingham	1,009	Liverpool	479
Bristol	397	London	8,170
Cardiff	296	Manchester	435
Edinburgh	439	Newcastle	259
Glasgow	684	Southampton	208

Note: Figures are from the 2011 census

Photos **D** and **E** show places with very different population densities. Photo **D** is a typical city scene with many buildings, plenty of activity and a lot of people. Photo **E** was taken in Scotland. It shows part of the Highlands, a beautiful but sparsely populated area in the north.

Can you think why one place is crowded whilst the other has very few people? What is the area like where you live – is it crowded or is it sparsely populated? Can you suggest why it has that population density?

Activities

1 Copy and complete these sentences.
   a A **population distribution** map shows ...
   b **Population density** tells us ...
   c **Densely populated** means that ...
   d **Sparsely populated** means that ...

2 Map **F** shows the spread of population in Britain.
   a Make a copy of the map and complete the key.
   b Write a paragraph to describe the distribution of population. Include the following words in your description:

- spread
- densely
- unevenly
- north and west
- south and east
- sparsely

3 a List the cities from table **C** in order of size. Give the biggest first.
   b Give three advantages of living in cities.

4 Look at sketch **B** and think carefully about what people need to live their everyday lives.
   a Study photo **D** and make a list of the things that would help people to live there.
   b Study photo **E** and suggest why very few people live in that area.

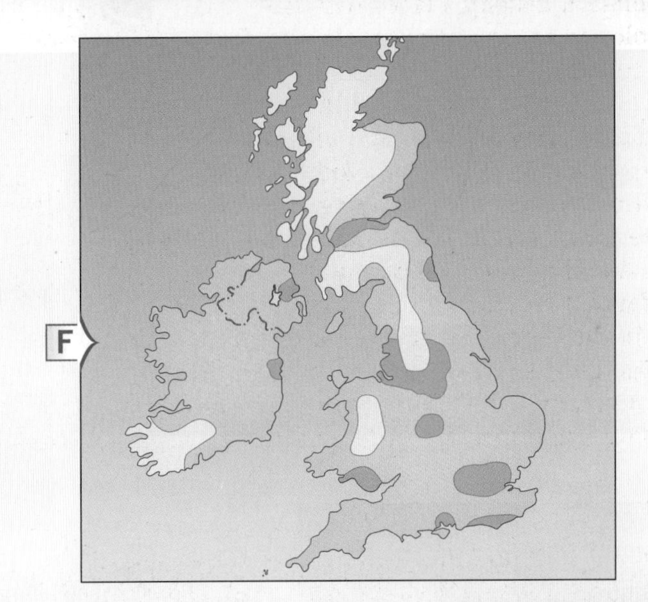

5 Draw a bar graph to show the number of people in each of the cities in table **C**.
- Arrange the bars in order of size with the biggest on the left.
- Use different colours for the cities in England, Scotland, Wales and Northern Ireland.
- Give your graph a title.

Summary

People are not spread evenly over Britain. Some areas are very crowded while others are almost empty. Population density is a measure of how crowded an area is.

What affects where we live?

Not only is the distribution of population uneven in Britain, but it is uneven throughout the world. There are now over 7 billion people in the world yet most of them live on only a third of the land surface. Like Britain, some areas are very crowded and others are almost empty.

There are many reasons for this. People do not like to live in places which are too wet or too dry, too hot or too cold. Nor do they like places that are mountainous, lack vegetation, are densely forested or liable to flood. People prefer pleasant places in which to live. They want to be able to earn money by working and have food available through farming or from shops. They like to be near to other people and have things to do and places to go.

Factors that discourage people from settling in an area are called **negative factors**. Factors that encourage people to live in an area are called **positive factors**. Some of these are shown in the photos below and in diagram **G** on the next page.

Look carefully at photos **A** to **F** and for each one in turn try to work out why it is likely to be either a densely populated area or a sparsely populated area.

A Himalayan mountains

B Amazon rainforest

C Western Europe

D Sahara Desert

E Polar regions – Antarctica

F Bangladesh

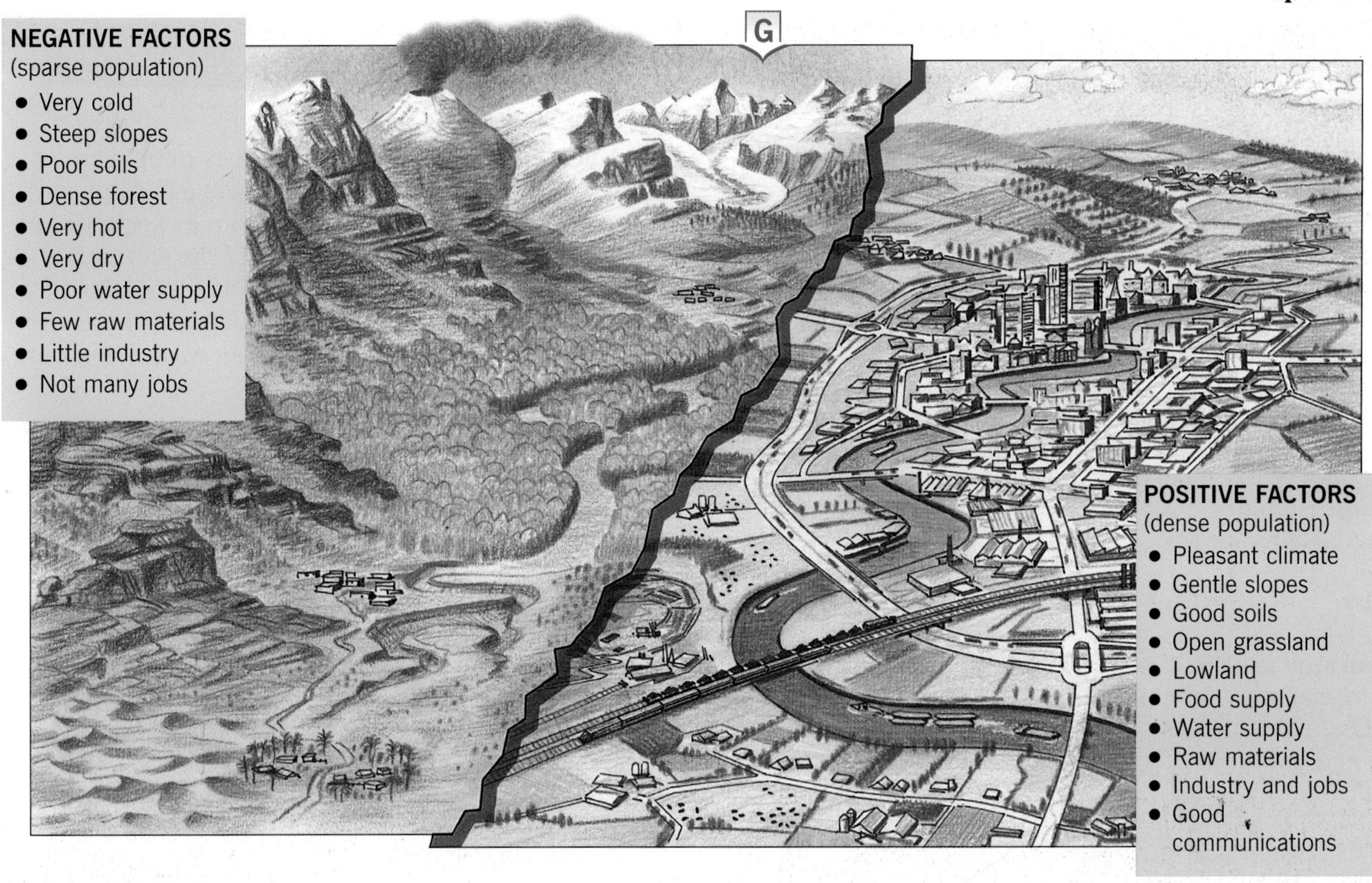

Activities

1 a Which one of the photos **A** to **F** does this list of words and phrases best describe?

• steep slopes • snowy • very cold • mountainous • icy • no soil • no industry • very few people

b Which of these could be used to describe photo **F**?

• dry • steep • level • poor soils • wet • good farming • factory work • cold • hot • sparse population • many people

c Imagine that you are passing through the desert in photo **D**. Make a list of words and phrases to describe what it would be like. Try to give at least **eight** different things.

2 Copy table **H** and put the factors from list **I** into the correct columns in your table.

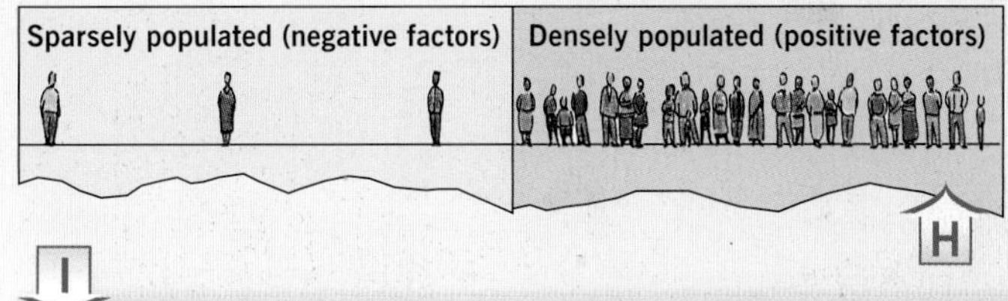

I

• flat land • mountains • dense forest • lowland • open grassland • good farming • unreliable water supply • deep, rich soils • thin, poor soils • job opportunities • poor farming • deserts

3 Give **two** reasons why few people live in:

a mountain areas

b desert areas.

4 Is the place where you live crowded or sparsely populated? What are the reasons for this? List the factors from diagram **G** which affect your area. Add any others that you think are important.

Summary

The way people are spread across the world is affected by many different things. These include relief, climate, vegetation, water supply, raw materials and employment opportunities.

Where do we live?

The photo-map below is quite remarkable. It is made up from more than 37 million satellite images carefully put together to give a picture of the world. The red dots have been added to show the distribution of population. Look carefully and you can see many of the world's major features. The cold polar regions show up as white. The densely forested parts of South America and Africa are a lush green. The areas that are dry and lacking vegetation are shades of brown. Can you see the great mountain ranges? They show up as patches or streaks of white.

The map also confirms how unevenly people are spread over the world. Vast areas have hardly any people living in them whilst other areas seem to be very crowded. Try to name some of the emptiest places. Places with a lot of people include parts of Western Europe, India, China and Japan. Where else in the world does the photo-map show that there are a lot of people?

Photos of the six areas described below are shown on page 52.

A

Amazon rainforest
Too hot and wet for people.
Dense forest makes communications and settlement difficult.
Sparsely populated.

Western Europe
Low-lying and gently sloping.
Pleasant climate.
Good water supply and soil for farming.
Easy communications and many resources for industry.
Densely populated.

Himalayan mountains
Too cold for people.
Steep slopes are bad for communications and settlement.
Poor, thin soil unsuitable for crops.
Sparsely populated.

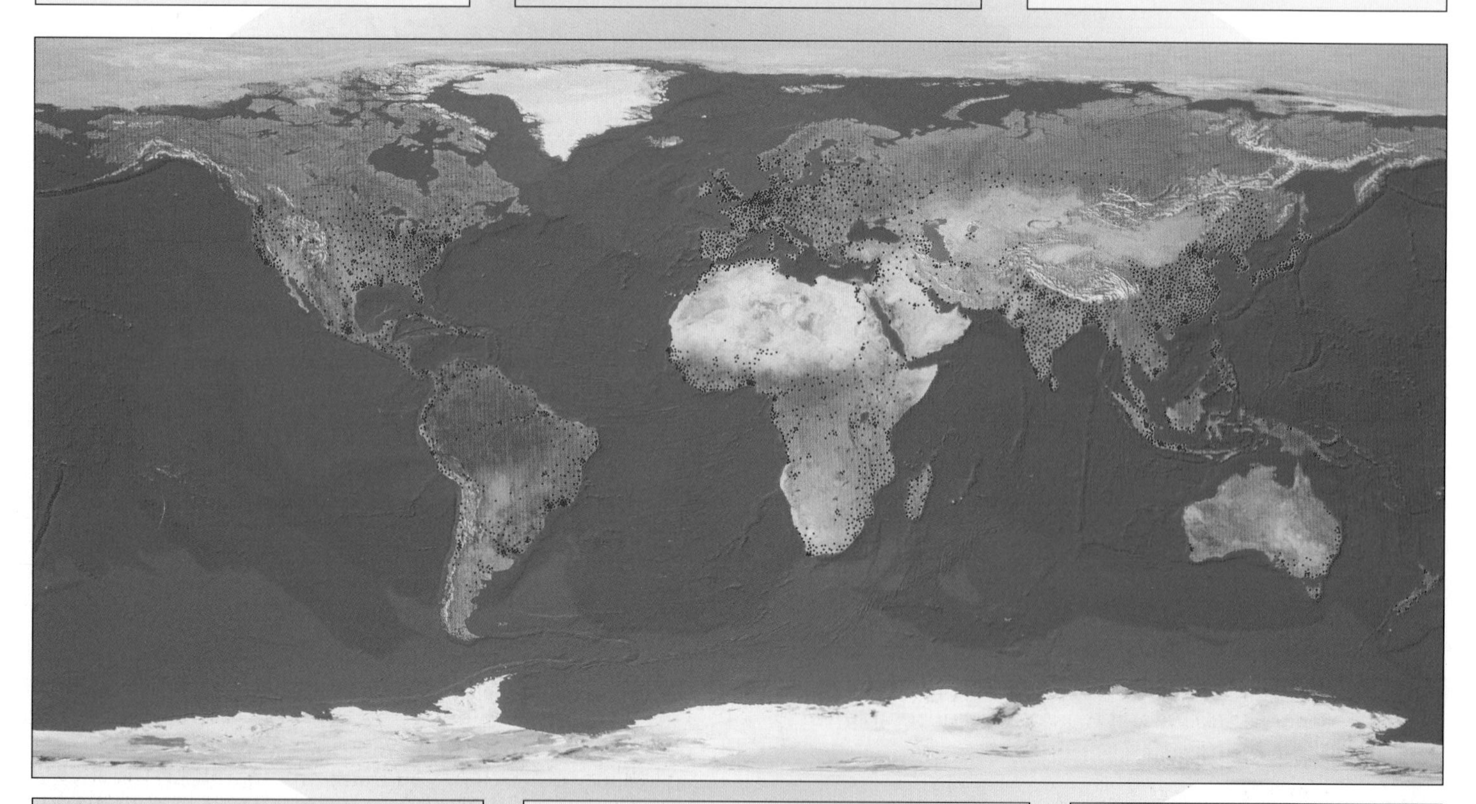

Polar regions – Antarctica
Too cold for people.
No soil for crops.
Snow and ice make communications and settlement very difficult.
Sparsely populated.

Sahara Desert
Too hot and dry for people.
Too dry and too little soil for crops to grow.
Sand makes communications difficult.
Sparsely populated.

Bangladesh
Low-lying and flat.
Rich, fertile soil. Hot and wet.
Ideal farming conditions.
Densely populated.

Cities are very popular places in which to live. They can provide housing, jobs, education, medical care and a better chance of getting on and enjoying life. More than half the world's population now live in cities and the number is increasing all the time.

The fastest-growing cities tend to be in the poorer countries. Here, the urban population is expected to double in the next ten years. This will produce some very large cities.

One of these, Mexico City, is expected to become one of the three largest cities in the world by 2020. By then it will have a population of almost 35 million. At present its population is increasing by nearly half a million people a year. That is the same as all the inhabitants of Liverpool or Edinburgh suddenly arriving in Mexico City in a single year. Think of the problems that such a rapid increase must cause.

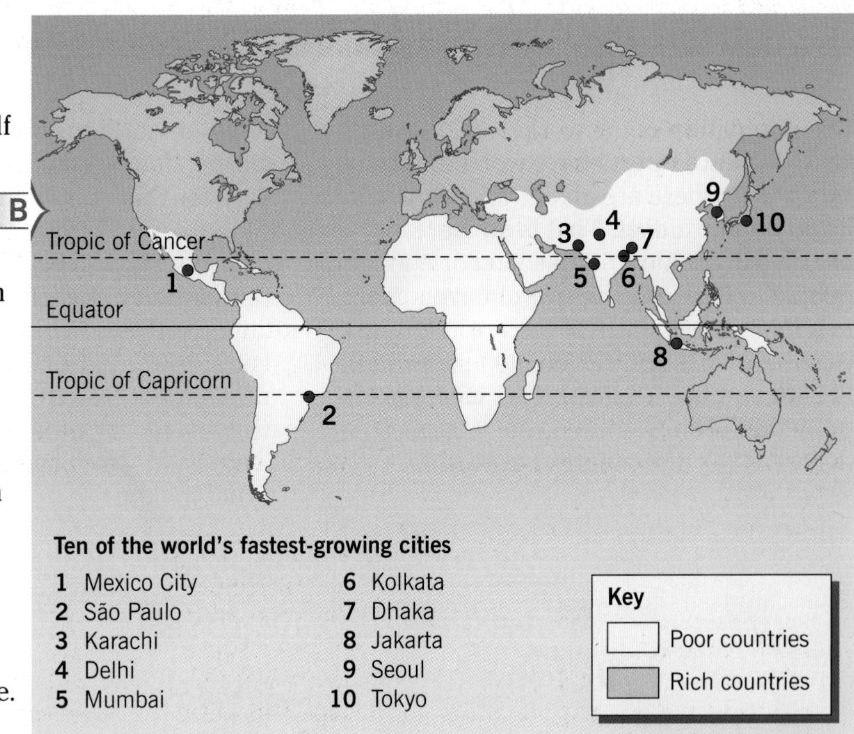

Activities

1 Copy and complete the sentences below using the following words:

- densely
- deserts
- uneven
- polar regions
- sparsely

a The distribution of population over the world is ______________ .

b The areas with the fewest people are the dense forests, ______________ and ______________ .

c Mountainous areas are ______________ populated.

d Areas with good resources and industry are ______________ populated.

2 Give named examples of:

a four densely populated areas

b six sparsely populated areas.

3 Which of the fastest-growing cities:

a are in South America

b are in Asia

c is in a rich country?

4 Copy diagram **C** and complete your diagram to show six reasons why people like to live in cities.

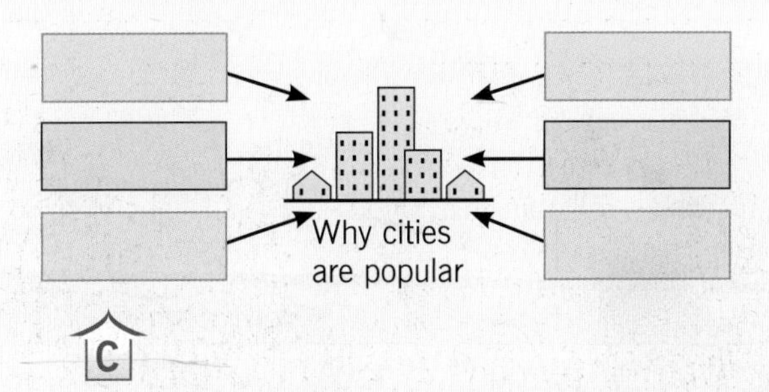

5 Of the eight statements given below, three are correct. Write out the correct ones. The ten fastest-growing cities are:

- mainly in poor countries
- mainly in rich countries
- mainly in polar regions
- mainly between the tropics
- on the coast
- spread all over the world
- in one continent
- in South America and Asia.

6 Use an atlas to name a country for each of the fastest-growing cities in map **B**.

7 With help from an atlas, try to find out why central Australia is sparsely populated and east and south-west USA are densely populated.

Summary

People are not spread evenly over the world. Some of the most crowded places are in China, India, parts of Western Europe, and some areas of Africa and the USA. More and more people in the world are living in cities.

How does population change?

The population of the world is increasing very quickly. Experts have worked out that every hour there are an extra 8,000 people living on our planet. That is an increase of about 2 people every second or enough people to fill a city the size of Birmingham in about a week. In 1987 the world's population passed the 5 billion mark and in 2012 reached 7 billion. This increase in population is now so fast that it is often described as a **population explosion**.

A major world problem is how to feed, clothe, house, educate, provide jobs and care for this rapidly increasing population.

Graph **A** shows the changes in world population between the years 1100 and 2012. Notice how the population increase is not even. Until 100 or 200 years ago population growth was actually very slow. Only in recent times has there been a real 'explosion'. Note that 1 billion is equal to 1,000 million.

A

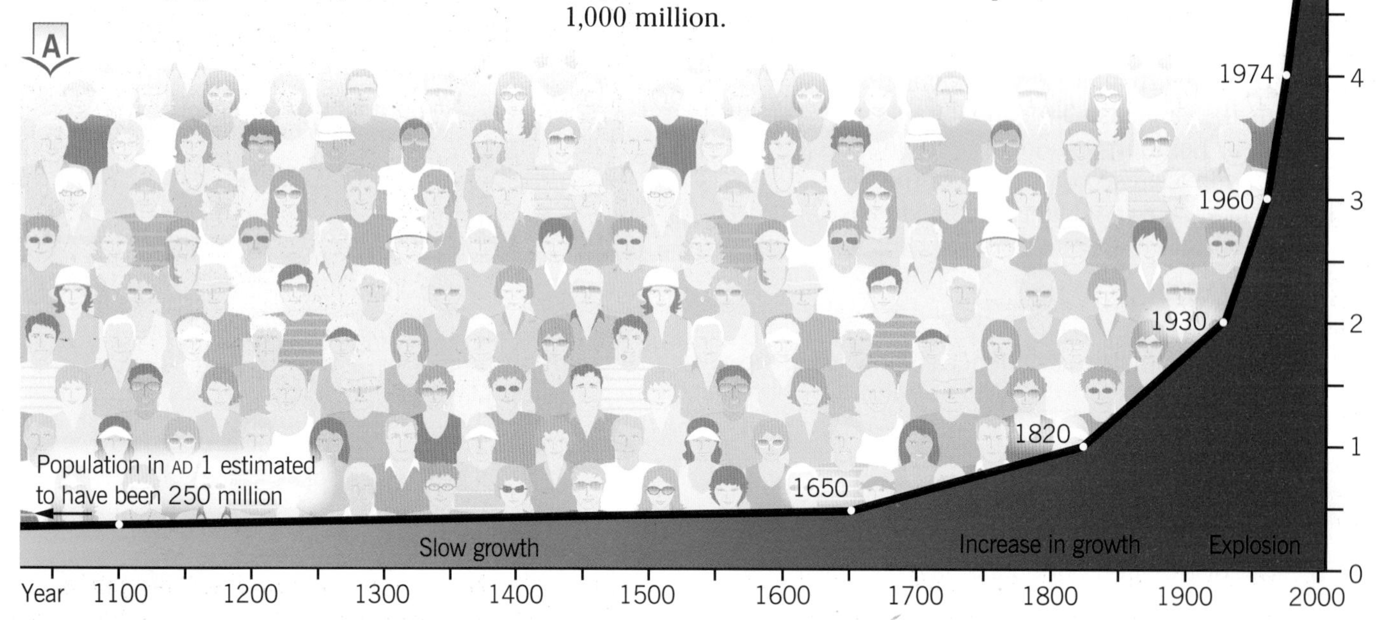

Population increases when the number of babies being born is greater than the number of people dying. The number of babies being born each year is called the **birth rate**. The number of people who die each year is called the **death rate**. Birth rates and death rates are measured as the number of births and deaths for each 1,000 of the population. The speed at which the population increases is called the **population growth rate**. Diagrams **B**, **C** and **D** show how the balance between births and deaths affects the population growth.

B

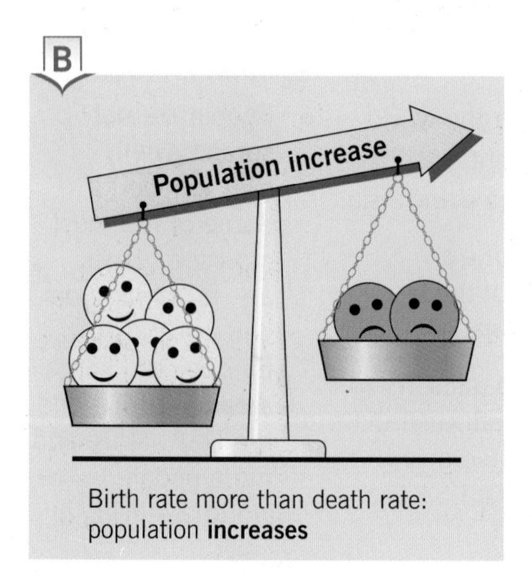

Birth rate more than death rate: population **increases**

C

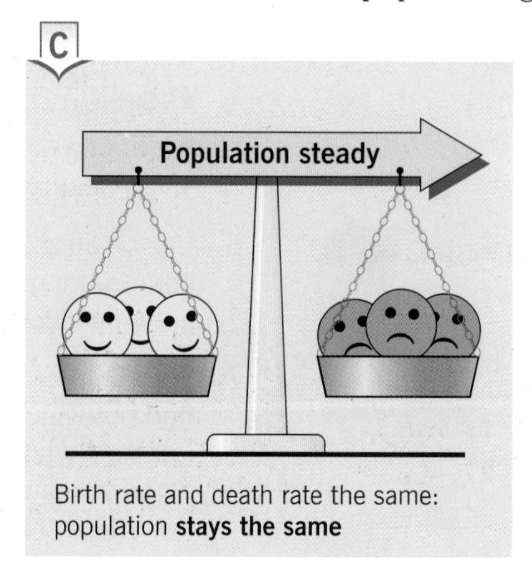

Birth rate and death rate the same: population **stays the same**

D

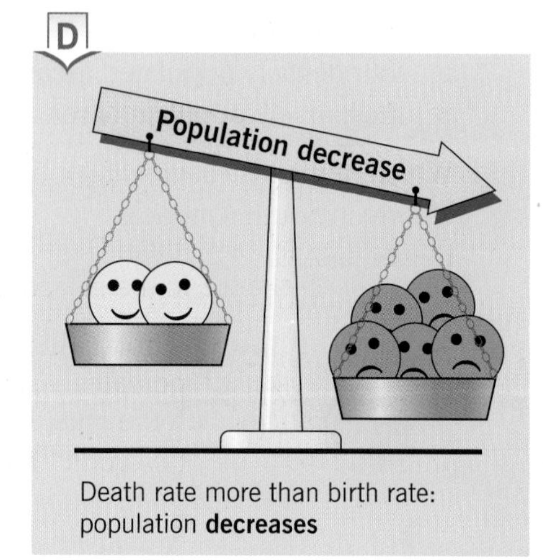

Death rate more than birth rate: population **decreases**

The population growth rate is not the same for all countries. In some, like the UK, the difference between birth and death rates is small so the population is changing only very slowly. In other countries, like Bangladesh, there are big differences between birth and death rates so the population is increasing rapidly.

Table **E** shows birth and death rates for some countries. Remember, the greater the difference between births and deaths, the larger the population change will be.

E

Country	Birth rate	Death rate	Natural increase
Bangladesh	20	6	14
Brazil	15	6	9
China	12	7	5
France	13	9	4
India	22	8	14
Italy	9	10	–1
Japan	8	10	–2
Mexico	19	5	14
UK	13	9	4
USA	13	8	5

- Figures given per 1,000 people. (World Bank 2012)
- Poorer countries are shaded yellow.
- **Natural increase** is the difference between birth and death rates.

F Things that can affect birth and death rates

Activities

1 a When did the world's population reach 1 billion?
  b How long did it take to double to 2 billion?
  c How long did it take to double again to 4 billion?

2 Describe the increase in world population shown in graph **A**.

3 a Write a sentence to explain what each of the following terms means:
  - birth rate
  - death rate
  - population growth rate.

  b Why is 'explosion' a good description of population changes since 1950?

4 Copy and complete table **G** by writing **increase**, **same** or **decrease** in the last column.

5 a List the countries from table **E** by the size of their natural increase. Put the ones with the greatest increase first.
  b What do you notice about the richer and the poorer countries?

6 Draw table **H** and from diagram **F** sort the things that affect birth rates and death rates into the correct columns.

H

Birth rate		Death rate	
High	Low	High	Low

7 Suggest **three** reasons why people in the UK may be more likely to live longer than people in poorer countries.

G

Births	Deaths	Population change
↗	→	
→	↗	
→	→	
↗	↗	

Summary

The world's population is increasing at a very rapid rate. Growth is very much faster in the poorer countries than in the richer ones. Population change in a country depends mainly on the birth and death rates.

What is migration?

London is the UK's largest city with a population of just over 8 million. After a period of decline, London's population is now showing a steady growth.

This growth is expected to continue for many years. This is partly because of natural increase but mainly because, like most cities, London acts like a magnet and attracts people from other places to live there.

People who move from one place to another to live are called **migrants**. They have a big effect on populations because they increase the numbers and can alter the mix of people who are living in a place.

Graph **A** shows London's growth since 1961. The photos in **C** show some of the people who have moved into London. Notice how different the people look and what varied backgrounds they have. Their comments help to explain some of the reasons why people migrate.

A

Population of London

Millions of people

9.0, 8.5, 8.0, 7.5, 7.0, 6.5

1961, 1981, 2001, 2021

Year

Estimated

B London

C

Hamish

I'm from a small village in Scotland and suffer from poor health. My family live in London so I've come to join them. It will also be easier to get medical care here.

Caroline

I went to university in Bristol but I'm ambitious and think that I'll have a better chance of getting a good job and enjoying life in London.

Janine

I qualified as a nurse in New York. I worked in Mexico for a while but have now moved to London to gain more experience and make new friends.

Chang

I came here from China. My company in Shanghai sent me to London some years ago and I like it so much that I'm going to stay.

Carmel

I've lived here all my life but my family are originally from Jamaica. They came here in the 1950s when the British government asked for people to fill job vacancies.

Zaric

I was brought up in a small town in Croatia. My family were killed in the war there and I always felt in danger. I moved to London to start my life again.

Migration is when people move home. The movement may be just around the corner to a better house. It may be from one part of the country to another in search of a job. It might be from one country to another for a different way of life. For many people country areas have very little to offer so they move to the towns and cities. This is called **rural-to-urban migration**. For other people a move to a different country holds many attractions. This is called **international migration**.

People migrate for two reasons. Firstly, they may wish to get away from things that they do not like. These are called **push factors** and include a shortage of jobs and poor living conditions. Secondly, people are attracted to things that they do like. These are called **pull factors** and include pleasant surroundings and good medical care.

Activities

1 What do each of these terms mean?
- Migration
- Rural-to-urban migration
- International migration

2 a Copy and complete bar graph **E** to show population growth in London.

b Describe the change in population since 1960.
- What is the overall change?
- Is the recent change slow, steady or rapid?

c Suggest reasons for the change.

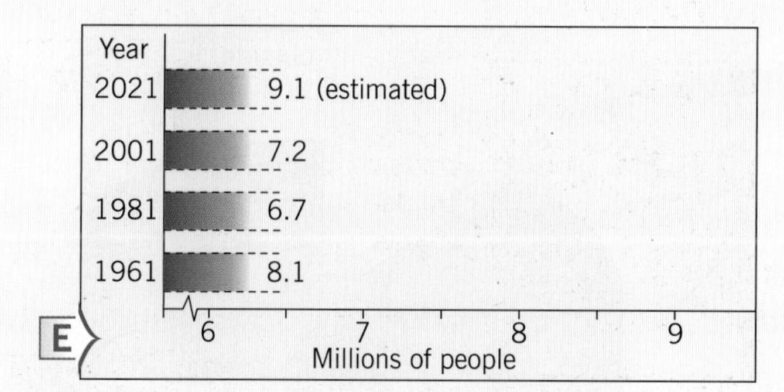

3 a What is a push factor? Give two examples.

b What is a pull factor? Give two examples.

4 Copy table **F** and use it to explain why the people in **C** migrated to London.

F

	Push factor	Pull factor
Hamish		
Janine		
Carmel		
Caroline		
Chang		
Zaric		

5 Imagine you are a migrant like Zaric in **C**.

a You have just arrived in London. Describe your first few days there. Think about language difficulties, getting a job, food, shelter, making friends ...

b Describe why you migrated from Croatia to London. Include push and pull factors.

Summary

Migration is the movement of people from one place to another. This movement may be the result of push and pull factors. The migration of people affects population size and the variety of different people in a place.

Who migrates to the UK?

Look at the welcome signs on poster **A**. They all mean the same but are in different languages. The poster is from Newham in London where many of the residents are newly arrived and have a language other than English.

In fact people have been arriving in Britain from other countries for more than 2,000 years. Some have come as invaders, some to escape problems in their own countries and some simply to find jobs and enjoy a better way of life.

Indeed the UK population is made up of **immigrants** and has always been a country of mixed races and cultures. The majority of UK residents are descended from Romans, Vikings, Angles, Saxons and Normans. The Irish have also settled in Britain for several centuries, while many other Europeans migrated here both during and after the Second World War.

Some of the largest groups of more recent immigrants are from countries that were once part of the British Empire, like India, Pakistan, Bangladesh and the Caribbean. After the war there were serious labour shortages in Britain. The government of the time invited people from these countries to come and fill job vacancies. Almost 1 million responded to the call. Most immigrants settled permanently in Britain. They had families and, along with their descendants, became UK citizens.

A

The graphs below show where people now living in the UK were born. More than 1 in 8 (7 million) were born overseas.

B

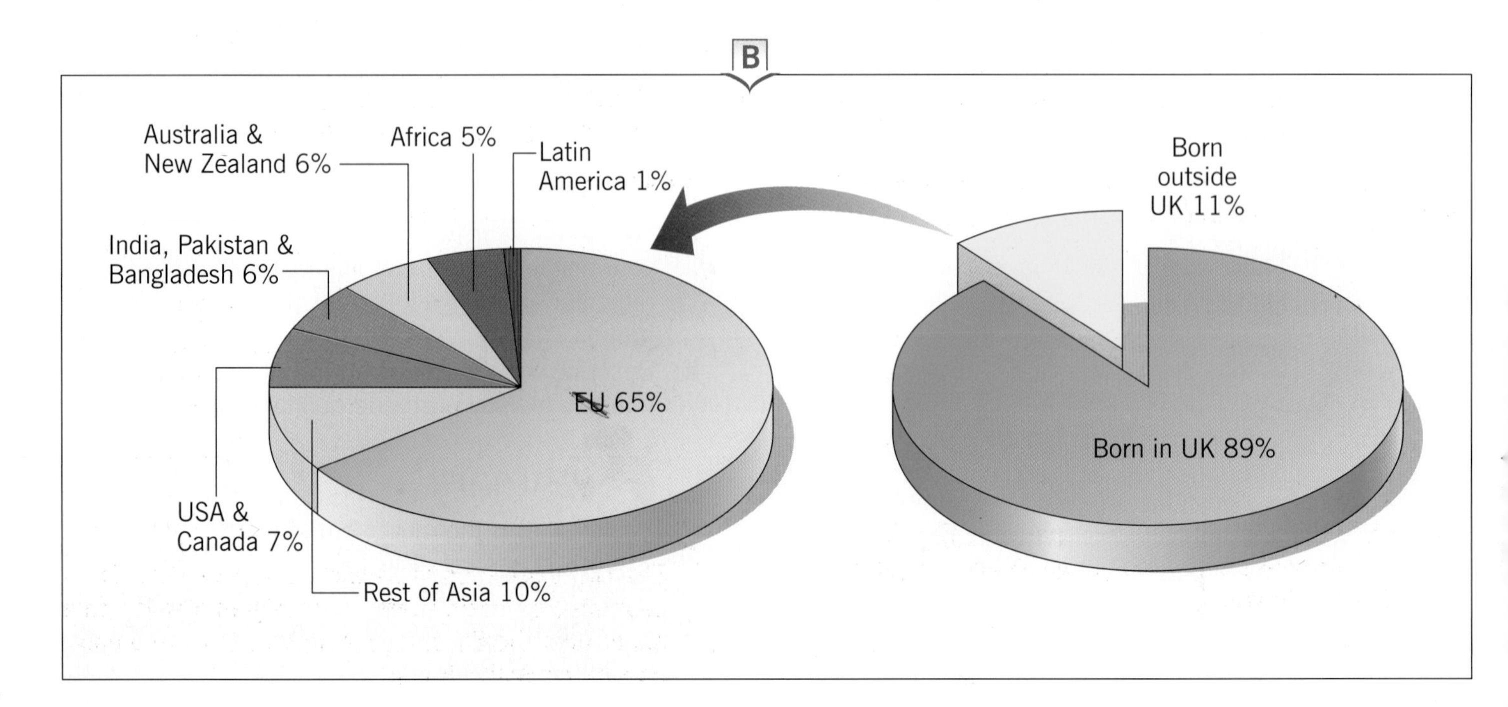

In recent years there has been a rapid increase in the number of migrants arriving in Britain. Most of these are from Eastern Europe. These countries joined the European Union between 2004 and 2014 and now have easier access to other countries in the EU. Many Eastern Europeans have left their home country in search of a better life elsewhere. Other migrants include **asylum seekers**. These are people who live in danger in their own country and want to move to a place where they will be safe.

Some people are against immigrants. They say that if numbers continue to increase, they will ruin the country. Others are in favour of them. They argue that most are genuine people with a need for a better life and a right to access the UK. They point out that most are well educated and are able to make a positive contribution to UK life.

There is concern, however, that there are just too many migrants coming into the country. Although these people generally bring benefits, if there are too many of them there can be problems. These problems are greatest in the big cities where a quarter of the people living there are immigrants and nearly one in ten homes has no one who speaks English as their main language. For this reason, the UK government is looking into ways of restricting and controlling the number of immigrants entering Britain.

C Asylum seekers at Calais

D

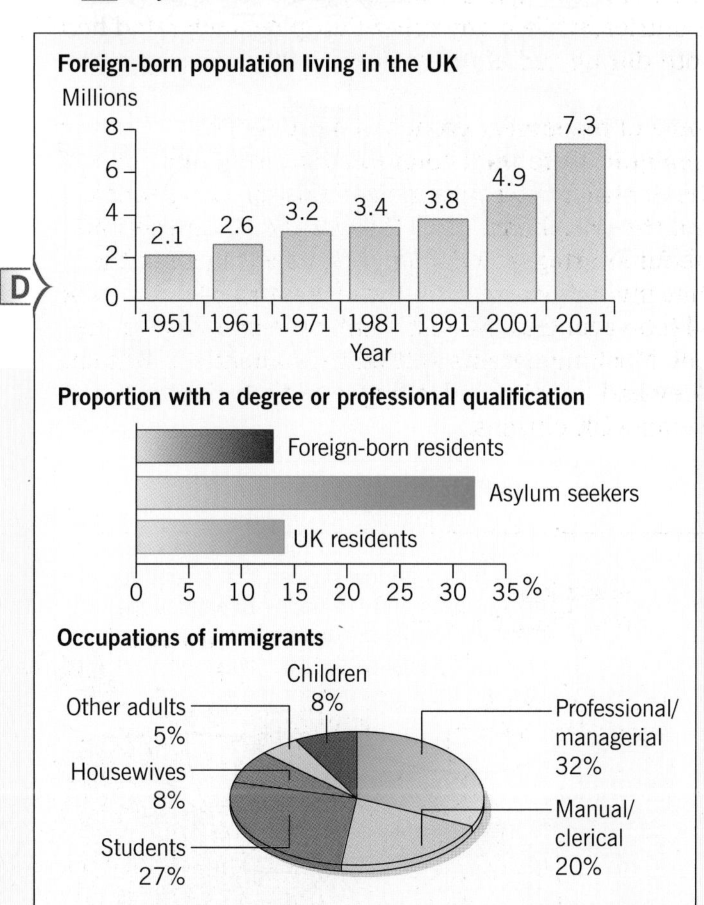

Activities

1. **a** How many different languages are shown on poster **A**?
   b Write out the welcome signs from three European countries and three non-European countries.
2. Make a list of where people in your class come from. Try to go back to previous generations. Sort the list under the headings on graph **B**.
3. Look at graph **B**.
   a Where have most immigrants come from?
   b Suggest why immigrants are allowed into the UK from this area.
   c Which groups of immigrants are likely to have relatives in the UK already?
4. Write a short letter to a newspaper supporting asylum seekers. Zaric on page 58 and the graphs in **D** will help you.

Summary

The UK is made up of people from many different countries. In recent years, most migrants have come either from the EU or from countries that were once part of the British Empire.

What are the effects of migration?

The effect of migration into Britain has been considerable. It has caused an increase in numbers and altered the mix of people in the country. It has also produced a **multicultural society** where people with different beliefs and traditions live and work together. Most people agree that this has added variety and interest to the UK.

There are many other effects of migration. Some affect the migrant and others affect the place they have moved from as well as the place they have moved to. Look at the comments in **A** and **B** which show some of these effects.

A **Different views on migration**

B

Activities

1 a Give five reasons why the Singh family are pleased they migrated to the UK.

  b Give five reasons why the Khafoor family are unhappy about migrating to the UK.

2 Look at the list of effects of migration in **C** below. Arrange the statements in a diamond shape as shown in drawing **D**.

- Put the advantage you think is the most important at the top.
- Put the next two below, and so on.
- The greatest disadvantage will be at the bottom.

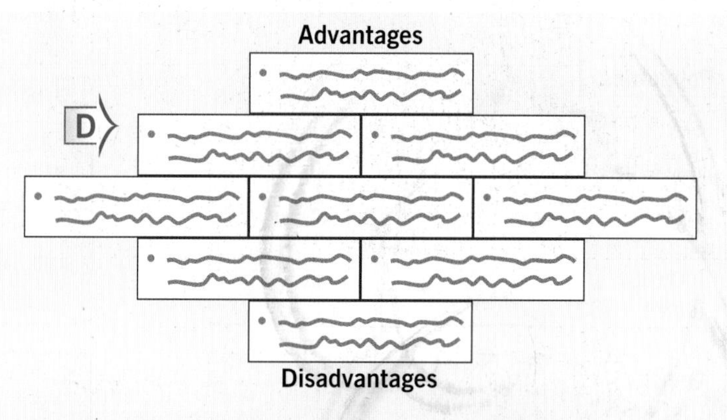

3 Suggest reasons for the headlines in **E** below. Answer in the form of a short newspaper article for each one.

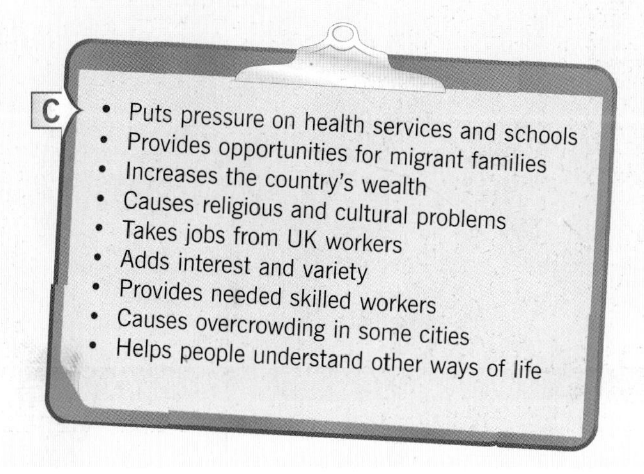
C
- Puts pressure on health services and schools
- Provides opportunities for migrant families
- Increases the country's wealth
- Causes religious and cultural problems
- Takes jobs from UK workers
- Adds interest and variety
- Provides needed skilled workers
- Causes overcrowding in some cities
- Helps people understand other ways of life

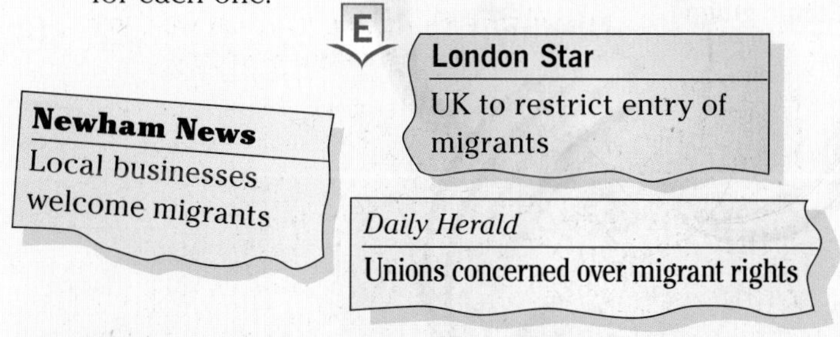
E

London Star
UK to restrict entry of migrants

Newham News
Local businesses welcome migrants

Daily Herald
Unions concerned over migrant rights

Summary

Many immigrants settle happily in their new surroundings but for some there can be difficulties. Migrants can be a great help to the economy. They provide much needed skills and add variety and interest to the UK.

How can we compare local areas?

This unit is about people: where they live, how many there are and how some of them move from one place to another. So what about the people in the area where you live? How many are there, how crowded is it and where do they come from? Learning about other places can also be interesting. What are they like and how do they compare with the place where you live?

Information about your local area can easily be found on the internet. The government's **National Statistics** website (www.statistics.gov.uk) gives facts and figures for the whole country. Typing in your postcode or town name will give you information about where you live. You can easily find out about other areas using the same method. You will also be given UK averages to compare with your own.

The website has statistics on a variety of population topics including population change, population density, age groups and ethnic origin.

It even lists the most popular names for babies! Much of the data is presented as graphs to make it clearer and help you make comparisons.

The information on these two pages is for the London boroughs of Newham and Chelsea. Notice how different they are. Which one is most like your local area? Which one would you prefer to live in?

A

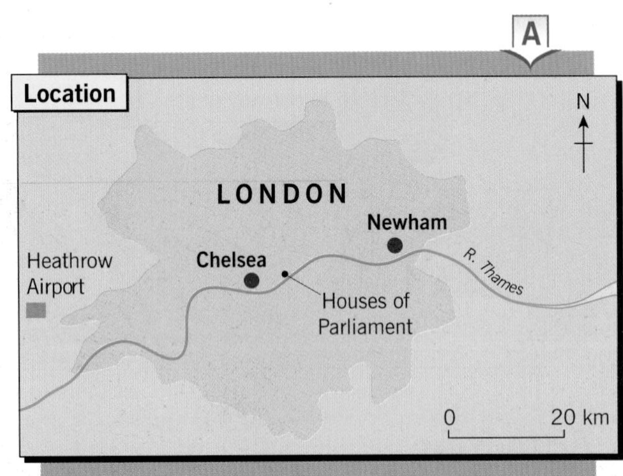

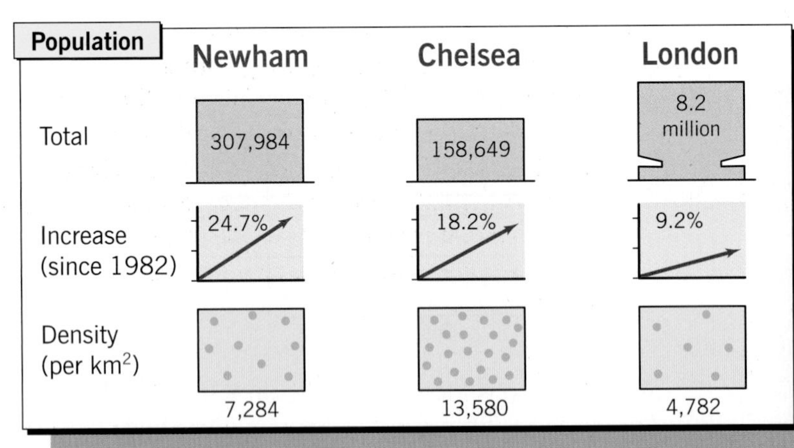

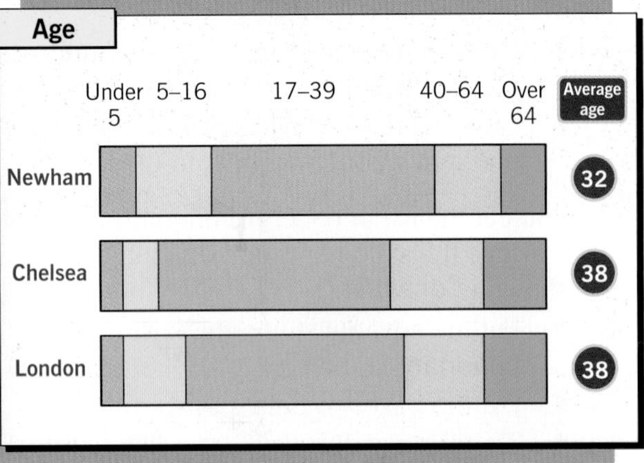

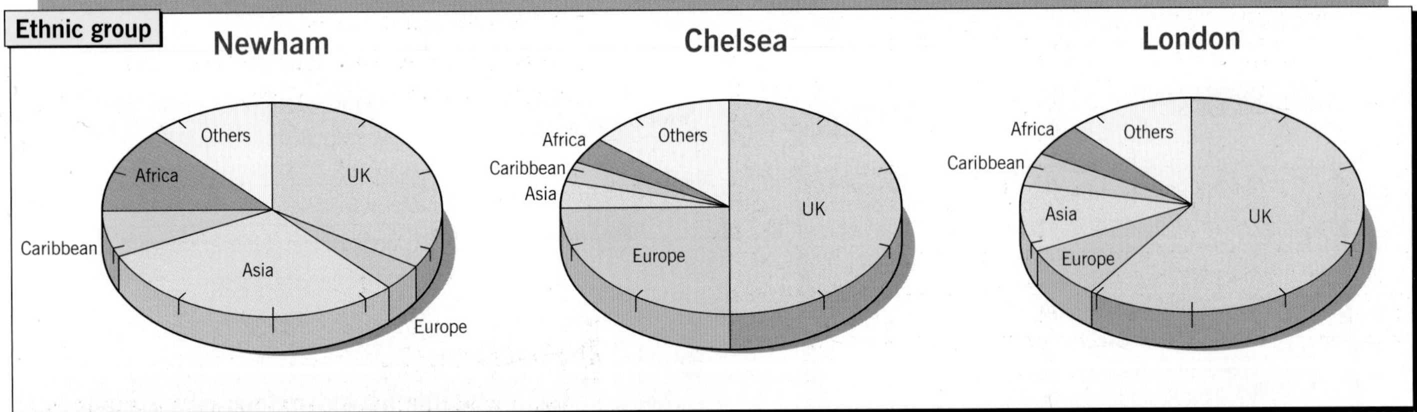

B

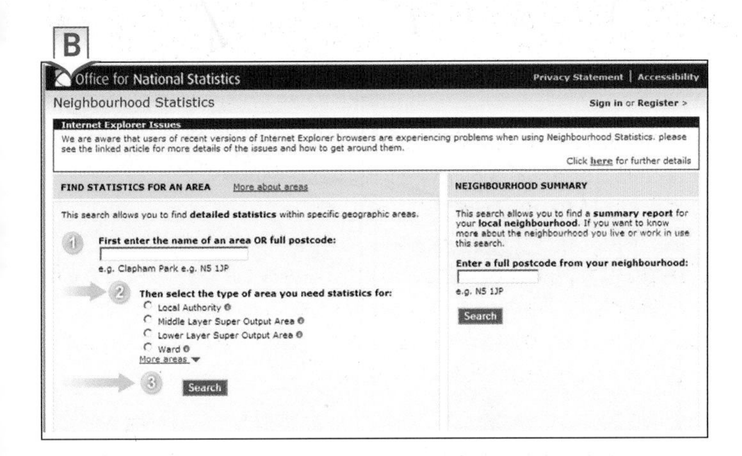

C Street market in Newham

D

	Newham	Chelsea
Closest to the city centre		
Has most people		
Has faster growth than London		
Is most crowded		
Has most young people		
Has most at retirement age		
Has more over 40s than London		
Has fewest from Europe		
Has greatest mix of nationalities		
Likely to attract more migrants		

Activities

1 a Make a larger copy of table **D**.
  b Tick the correct column for each statement. More than one column may be ticked.

2 a Copy table **E** below and complete it using information from the 'Ethnic group' graph in **A**. The first item has been done for you.
  b Look at your completed table and describe the main differences between Newham and London as a whole.

E

	Ethnic groups	
	Newham	**London**
Highest (%)	1 *UK (34%)*	1
	2	2
	3	3
	4	4
	5	5
Lowest (%)	6	6

3 You work for the council and have been asked to produce a report comparing your local area with the UK. Your report is about the population and people who live in the area.
  a Go to the Neighbourhood Statistics website www.neighbourhood.statistics.gov.uk. You should see a page like the one in **B**.
  b Enter your postcode or town name.
  c Click 'Local authority'.
  d Click 'Search'.
  e Now collect information for your report.

Your report should present some information as graphs, maps and tables.
Remember to explain what it shows and compare it with the UK averages.

Summary

Information about population and people may be found on the internet. This can be used to compare your local area with other places.

The population enquiry

As we have seen earlier in this unit, people are not spread evenly around the world. Some places are crowded whilst other places have very few people. There are many reasons for this. Some are **positive factors** which encourage people to settle in an area. Others are **negative factors** which discourage settlement. There are almost always some positive factors and some negative factors working in an area. It is the balance between the two that determines how crowded a place becomes.

Look at drawing **D** on the next page which shows an imaginary continent in the northern hemisphere. It shows many of the factors affecting population distribution that were described on pages 50 to 55. Your task in this enquiry is to determine the most likely population distribution on the continent using negative and positive factors.

A

How can positive and negative factors affect where people live?

1 **a** Make a larger copy of table **B**.
b List the positive and negative factors for each of the areas labelled 1 to 10 on drawing **D**.
c For each place, decide if it is likely to be crowded, to have few people, or to be in between. Write your decision in the 'Population density' column.
d In the last column of the table, give reasons for your suggested population density.

2 **a** Make a larger copy of map **C**.
b Show the population distribution of the continent by dividing it into areas that are likely to be crowded, to have few people, or to be in between. Use colours similar to those on map **A** on page 50.
c Choose the most likely locations of three main towns and mark them on your map.
d Complete your map by colouring the coastline blue and giving names to the towns, mountain ranges, rivers, seas and imaginary continent.

3 Describe and suggest reasons for the locations that you have chosen for the three main towns.

B

Area	Positive factors	Negative factors	Population density	Reasons
1				
2				
3				
4				

C

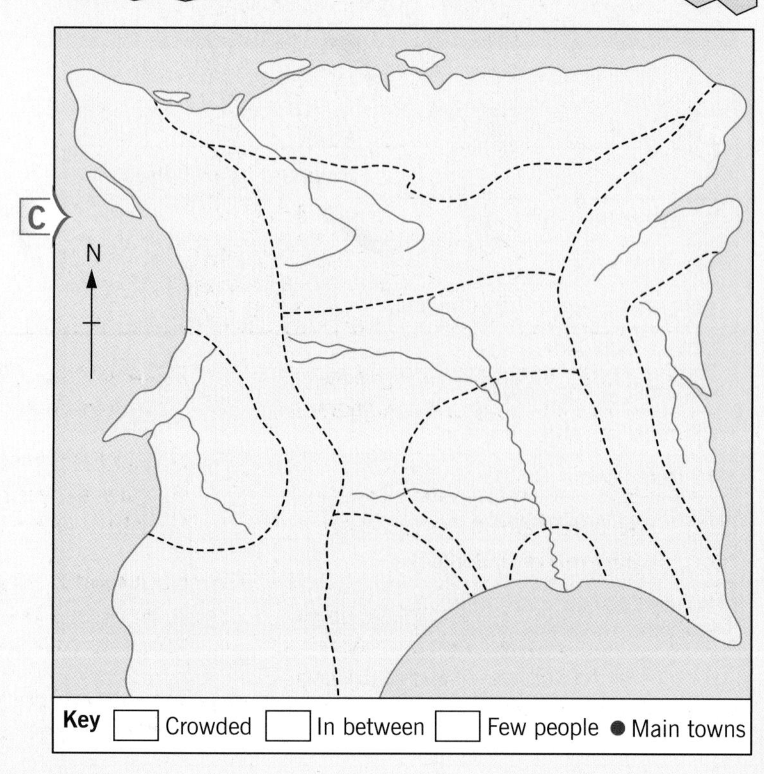

D

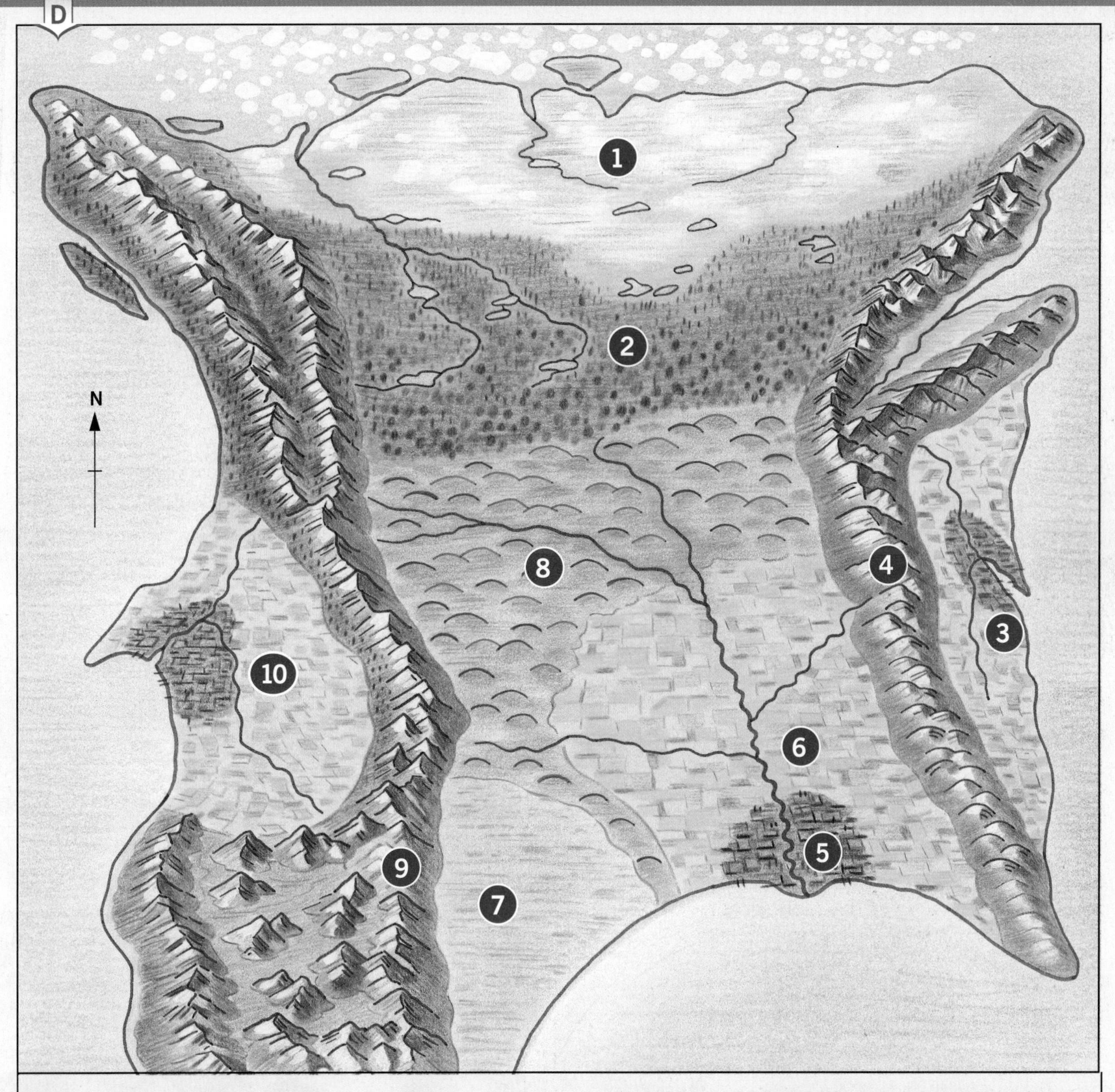

1. Mainly flat and low-lying land. Poor, thin soils. Little vegetation. Very cold.
2. Hilly with many lakes. Poor soils. Dense coniferous forest. Very cold winters.
3. Flat or gently sloping land. Raw materials available for industry. Fertile soils. Warm climate with plenty of rain.
4. Steep slopes. Thin soils. Few jobs. Poor communications.
5. Flat and low-lying land. Serves rich farming area inland. Industries and port facilities.
6. Flat or gently sloping land. Good-quality farmland. Warm summers with plenty of rain.
7. Very hot and dry. Mainly desert with few rivers. Little agriculture.
8. Mainly hilly with poor soils. Large-scale cattle farming. Few jobs. Cold winters.
9. Rugged, mountainous area. Difficult communications. Few jobs. Poor, thin soils. Wet.
10. Flat and low-lying. Fertile soils and good farming. Industries and port facilities. Good climate.

India and Asia

4 What is India like?

What is this unit about?

After a general introduction on Asia, this unit is about India, a developing country that in the last few years has been developing very quickly. It has extreme differences in relief, climate, population, standards of living and quality of life.

In this unit you will learn about:

- Asia's main features
- India's main features
- some of the contrasts in India
- its main physical features
- population distribution and movement
- differences between urban and rural life
- India's interdependence and development.

Why is learning about India important?

Learning about India helps you understand another country that is very different from the UK. The people are different too, although India is the mother country of many now living in Britain.

This unit can help you:

- broaden your knowledge of the world
- learn about different landscapes and climates
- understand ways of life that are different from your own
- recognise differences in a country
- understand differences in development
- develop an interest in other countries.

A Hindu pilgrims on the banks of the River Ganges, at Varanasi

How are photos **A**, **B**, **C** and **D** different to scenes found in your local area?

Think of differences, where appropriate, in the:

- number of people and the clothes they are wearing
- types of buildings
- weather, vegetation and wildlife.

B Religious festival in India

C Bengal tiger

D Taj Mahal

What are Asia's main physical features?

Stretching from the frozen Arctic to the hot Equator, Asia is by far the world's largest continent. Much of the land is barren, with vast, empty deserts and remote, treeless plains. The Himalayan mountains form a massive barrier between the heat of India to the south and the cold of Tibet to the north. The range is permanently snow-capped. The world's highest peak, Mount Everest, is located there.

Asia also has some of the world's most fertile plains and valleys, particularly along the great rivers such as the Mekong, Yangtze, Indus and Ganges. Elsewhere, the vegetation is hugely varied. In the north there is cold, frozen tundra and dense coniferous forest. In the south-east, the land is mainly mountainous or covered in tropical rainforest that are teeming with wildlife.

A Guilin, China

B Mt Everest, Himalayas

C Wadi Rum, Jordan

Activities

Refer to map **E** for Activities 1–3.

1 Use the map scale to measure the following:
   - a the length of Asia from **A** to **B**
   - b the width of Asia from **C** to **D**
   - c the length of the Yangtze, Ganges and Mekong rivers
   - d the length of the Himalayas.

2 Give the latitude and longitude of each of the following:
   - a Lake Baikal
   - b the mouth of the Ganges
   - c the Himalayas
   - d the Gobi Desert.

3 Describe Asia's physical features using the headings in diagram **D**. Write a few short statements for each one.

D **Asia: physical features**
- Earthquakes and volcanic activity
- Vegetation features
- River features

4 Write short descriptions of photos **A**, **B** and **C**. Pages 110 and 111 in *Foundations* will help you.

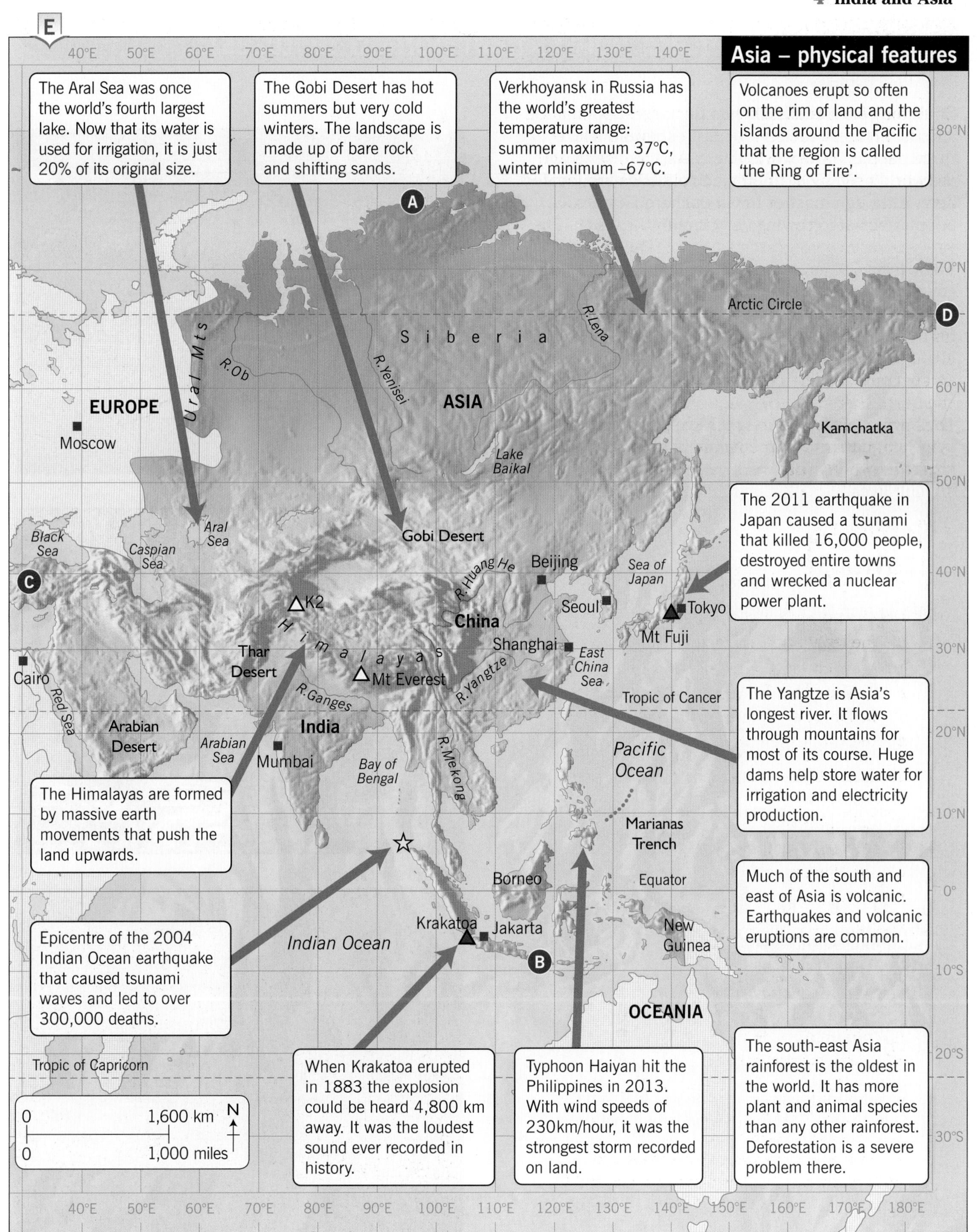
E
Asia – physical features
40°E
50°E
60°E
70°E
80°E
90°E
100°E
110°E
120°E
130°E
140°E
The Aral Sea was once the world's fourth largest lake. Now that its water is used for irrigation, it is just 20% of its original size.
The Gobi Desert has hot summers but very cold winters. The landscape is made up of bare rock and shifting sands.
Verkhoyansk in Russia has the world's greatest temperature range: summer maximum 37°C, winter minimum –67°C.
Volcanoes erupt so often on the rim of land and the islands around the Pacific that the region is called 'the Ring of Fire'.
80°N
A
70°N
Arctic Circle
D
R.Lena
Siberia
Ural Mts
R.Ob
R.Yenisei
60°N
ASIA
EUROPE
Kamchatka
Moscow
Lake Baikal
50°N
The 2011 earthquake in Japan caused a tsunami that killed 16,000 people, destroyed entire towns and wrecked a nuclear power plant.
Aral Sea
Gobi Desert
Black Sea
Caspian Sea
R.Huang He
Beijing
Sea of Japan
40°N
C
K2
Seoul
Tokyo
China
Himalayas
Mt Fuji
Thar Desert
Shanghai
East China Sea
30°N
Cairo
Mt Everest
R.Yangtze
R.Ganges
Tropic of Cancer
The Yangtze is Asia's longest river. It flows through mountains for most of its course. Huge dams help store water for irrigation and electricity production.
Red Sea
Arabian Desert
India
20°N
Arabian Sea
Mumbai
Bay of Bengal
R.Mekong
Pacific Ocean
The Himalayas are formed by massive earth movements that push the land upwards.
Marianas Trench
10°N
Borneo
Equator
0°
Much of the south and east of Asia is volcanic. Earthquakes and volcanic eruptions are common.
Krakatoa
Jakarta
New Guinea
Indian Ocean
Epicentre of the 2004 Indian Ocean earthquake that caused tsunami waves and led to over 300,000 deaths.
B
10°S
OCEANIA
The south-east Asia rainforest is the oldest in the world. It has more plant and animal species than any other rainforest. Deforestation is a severe problem there.
20°S
Tropic of Capricorn
When Krakatoa erupted in 1883 the explosion could be heard 4,800 km away. It was the loudest sound ever recorded in history.
Typhoon Haiyan hit the Philippines in 2013. With wind speeds of 230km/hour, it was the strongest storm recorded on land.
0
1,600 km
0
1,000 miles
N
30°S
150°E
160°E
170°E
180°E

What are Asia's main human features?

Of all the continents, Asia has the largest population and the greatest variety of cultures. Two out of every three people in the world live in Asia and seven of the world's ten most populated countries are located here. Although most of the people are farmers, city populations are growing very rapidly.

Apart from China, Japan and Thailand, most Asian countries were once colonies that were controlled by European powers. Gradually, these countries have become independent and developed their own ways of life. Many have enjoyed rapid economic growth and become modern, highly developed countries.

These changes have brought huge improvements in living standards for large numbers of people. Many others, however, still live in poverty and lead very difficult lives.

A Potola Palace, Lhasa, Tibet

B Dhaka, Bangladesh

C Pudong skyline, China

Activities

Refer to map **D** for Activities 1–3.

1 Give the latitude and longitude of each of the following cities:
  - a Beijing b Tokyo c Bangkok
  - d Delhi e Irkutsk f Singapore.

2 Name the countries that have a border with:
  - a India b China c Thailand.

3 Describe Asia's human features using the headings below. Write a few short statements for each one.

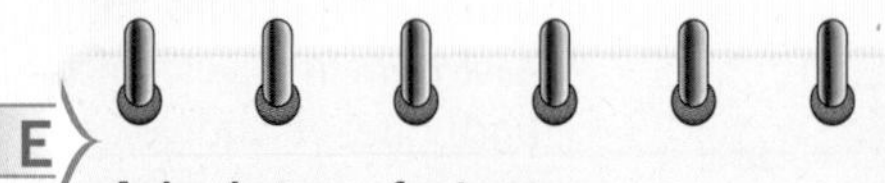

E

Asia: human features

- Population
- Industry and resources
- Growth and wealth
- Problems

4 Write short descriptions of photos **A**, **B** and **C**. Pages 110 and 111 in *Foundations* will help you.

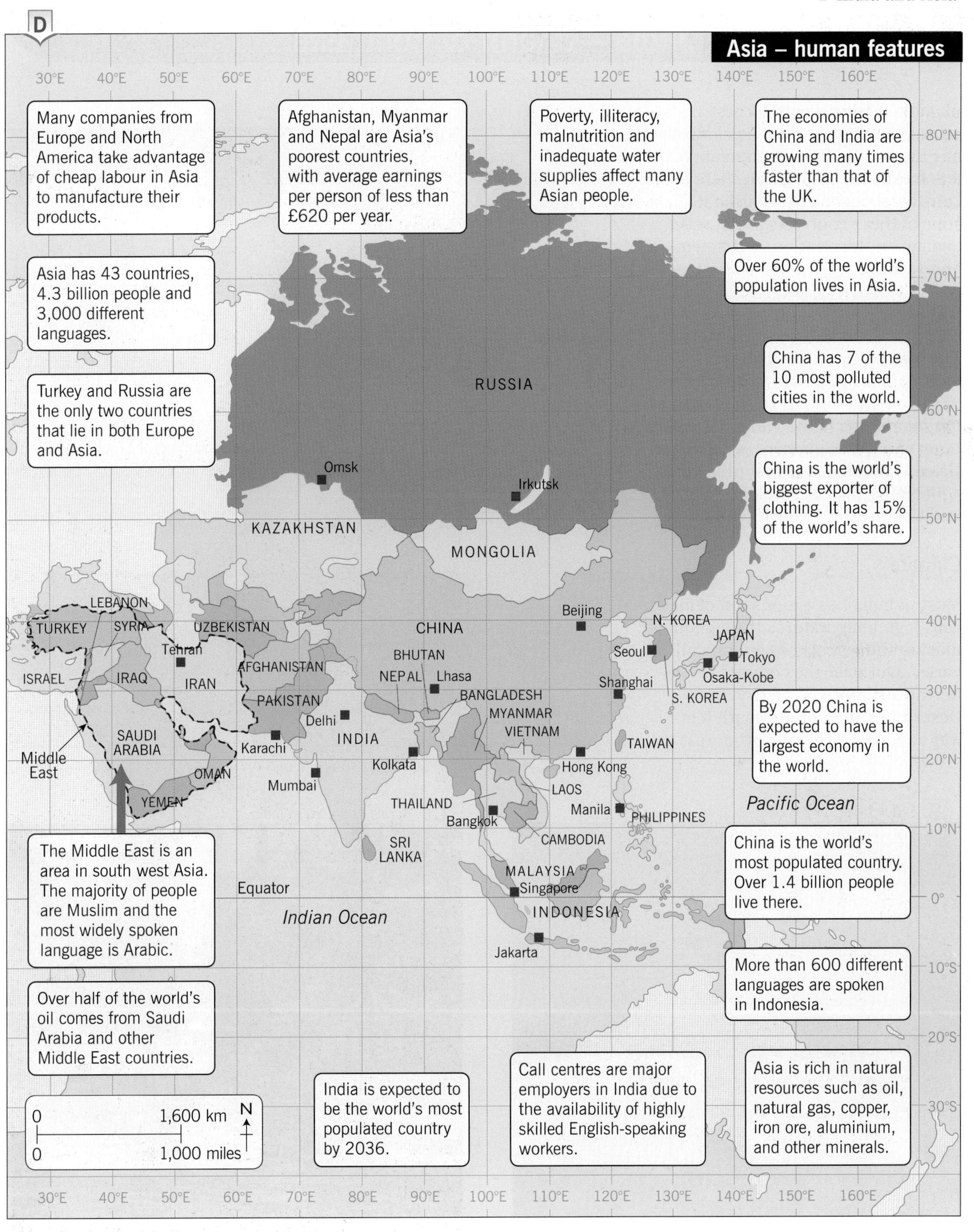
D
Asia – human features
30°E
40°E
50°E
60°E
70°E
80°E
90°E
100°E
110°E
120°E
130°E
140°E
150°E
160°E
80°N
70°N
60°N
50°N
40°N
30°N
20°N
10°N
0°
10°S
20°S
30°S
Many companies from Europe and North America take advantage of cheap labour in Asia to manufacture their products.
Afghanistan, Myanmar and Nepal are Asia's poorest countries, with average earnings per person of less than £620 per year.
Poverty, illiteracy, malnutrition and inadequate water supplies affect many Asian people.
The economies of China and India are growing many times faster than that of the UK.
Asia has 43 countries, 4.3 billion people and 3,000 different languages.
Over 60% of the world's population lives in Asia.
China has 7 of the 10 most polluted cities in the world.
Turkey and Russia are the only two countries that lie in both Europe and Asia.
China is the world's biggest exporter of clothing. It has 15% of the world's share.
RUSSIA
Omsk
Irkutsk
KAZAKHSTAN
MONGOLIA
LEBANON
TURKEY
SYRIA
UZBEKISTAN
CHINA
Beijing
N. KOREA
JAPAN
Tokyo
Seoul
Osaka-Kobe
Tehran
ISRAEL
IRAQ
IRAN
AFGHANISTAN
BHUTAN
NEPAL
Lhasa
Shanghai
S. KOREA
PAKISTAN
BANGLADESH
MYANMAR
Delhi
VIETNAM
TAIWAN
SAUDI ARABIA
INDIA
Karachi
Kolkata
Hong Kong
Middle East
OMAN
Mumbai
LAOS
YEMEN
THAILAND
Manila
PHILIPPINES
Bangkok
CAMBODIA
SRI LANKA
MALAYSIA
Singapore
INDONESIA
Jakarta
Equator
Indian Ocean
Pacific Ocean
By 2020 China is expected to have the largest economy in the world.
The Middle East is an area in south west Asia. The majority of people are Muslim and the most widely spoken language is Arabic.
China is the world's most populated country. Over 1.4 billion people live there.
More than 600 different languages are spoken in Indonesia.
Over half of the world's oil comes from Saudi Arabia and other Middle East countries.
0
1,600 km
0
1,000 miles
N
India is expected to be the world's most populated country by 2036.
Call centres are major employers in India due to the availability of highly skilled English-speaking workers.
Asia is rich in natural resources such as oil, natural gas, copper, iron ore, aluminium, and other minerals.

India – a land of contrasts

As India is the seventh largest country in the world by size and has the second largest population, it is hardly surprising that there are considerable contrasts within it. Four of these contrasts are in relief, climate and vegetation, settlement, and wealth.

Relief

The Himalayas is the highest mountain range in the world. It lies to the north of India. To the south of the mountains is the wide, flat valley floor and the huge delta of the River Ganges. Much of the rest of India is a plateau. This rises behind Mumbai and Goa as the Western Ghats.

Climate

Most of India has a **monsoon** climate with high temperatures throughout most of the year. However, rainfall varies a lot over the country. In the north-west there is the dry Thar Desert. In Kerala in the south it is wet enough for tropical rainforest to grow.

A Physical map of India

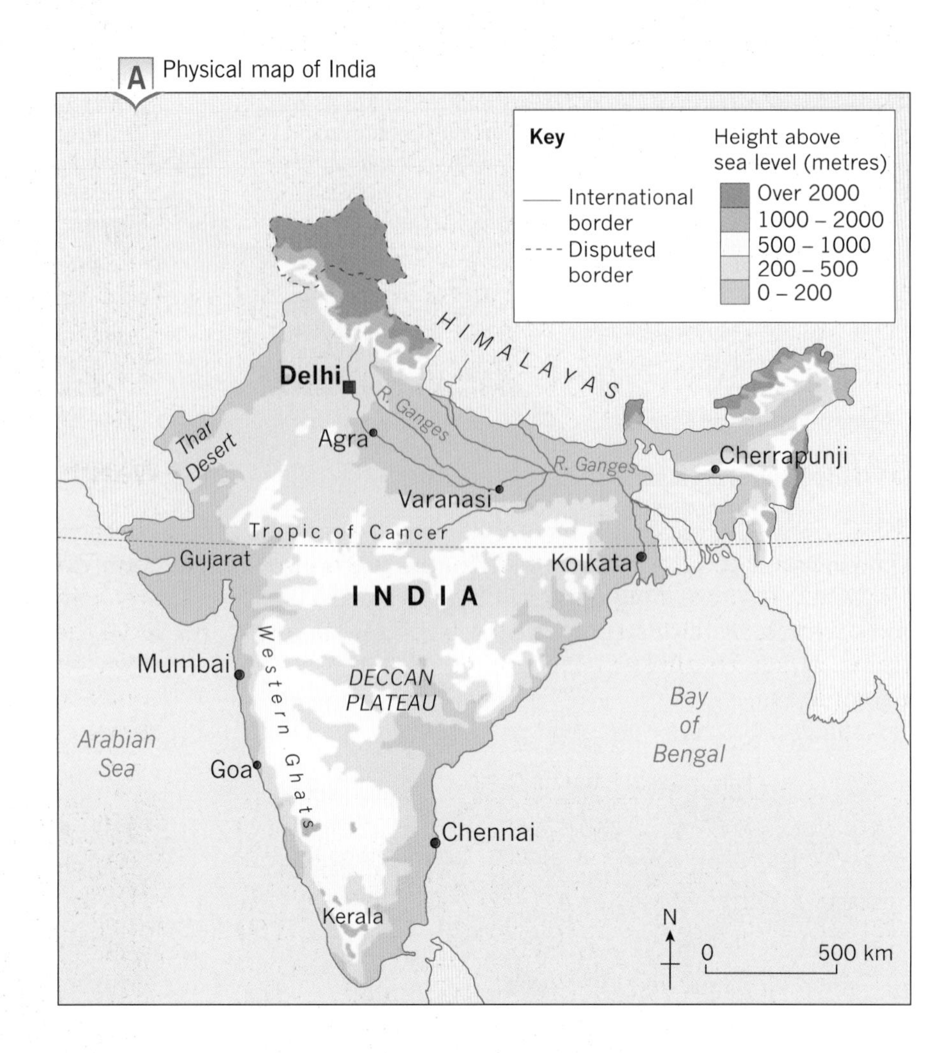

B Rice fields on the Ganges delta

C Kerala in southern India

D Village street in rural Kerala

E Main street in urban Delhi

Settlement

Many people in India still live in villages, although these are much larger than those found in Britain. However, an increasing number of people are moving to large cities such as Mumbai, Kolkata and Delhi, which are three of the world's ten largest urban areas.

Wealth and poverty

There is always a wide gap in wealth in developing countries. This can be between rural and urban areas and between the rich and the poor. In India the gap is greatest in the large cities. Here skyscraper office blocks and high-rise luxury flats contrast with poor-quality housing in Mumbai's **chawls** (see page 81) and Kolkata's **bustees**.

F Luxury flats in Uttar Pradesh, Delhi

G Slum huts in Kolkata

Activities

1 Describe the contrasts in:
   a relief between the Himalayas (photo **B** page 70) and the Ganges valley (photo **B** page 74)
   b climate between north-west India and the south
   c a street scene in a rural and an urban area (photos **D** and **E**)
   d wealth, as shown by housing in a large Indian city (photos **F** and **G**).

2 Using pages 68–69 and 74–75, find out where in India the following are located:

a	bustees	b	the Taj Mahal
c	chawls	d	Varanasi
e	tourist beaches	f	a delta
g	tropical rainforest	h	a hot desert
h	Kerala	i	snow-covered mountains.

Summary

India is a country of great contrasts. These include differences in relief, climate and vegetation, settlement and wealth.

What are India's main physical features?

You may already have heard of the term **plate tectonics**. It describes how the surface of the earth (crust) is divided into segments called plates. Plates float on molten magma (rock) that is found under the crust. The result is that these plates can move in different directions.

Two hundred million years ago, the plate with India on it was located next to southern Africa in the southern hemisphere. It then moved northwards until it finally collided with Asia. The impact of this collision caused the rocks between them to buckle upwards to form the high **fold mountains** that are the Himalayas (photo A). As India is still moving northwards, the Himalayas are still growing in height.

Unfortunately this movement is uneven. It is as if you (India) came across a larger, heavier object (Asia) in your path: you may have to push harder and harder until the object suddenly gives way. This sudden movement in the earth's crust is known as an **earthquake**. There are many earthquakes in the Himalayas. If they occur where people live they can cause great loss of life and damage to property. In 2001, several towns were destroyed and over 30,000 people were killed by an earthquake in Gujarat, in northern India.

A The Himalayas

India has a **monsoon** climate. The word monsoon means 'a season'. As graph C shows, India has two seasons: the south-west monsoon (June to October) and the north-east monsoon (November to May).

B After the Gujarat earthquake, 2001

During the south-west monsoon, between June and October, winds blow from the warm Indian Ocean. On their way they collect lots of moisture. When they reach the Western Ghats they are forced to rise. As they rise they cool to give very heavy **relief rainfall**. In an average year, Mumbai receives over 2,000 mm in five months (London gets 500 mm in a whole year).

As the winds continue towards the Himalayas they give even more rain as they are forced to rise even higher. Cherrapunji is said to be the world's wettest place: on average it gets 14,000 mm a year (see map **A** on page 74).

During the north-east monsoon, from November to May, the winds change direction and blow from central Asia. As this area is dry, the winds are unable to pick up much moisture, so India gets very little rainfall. Mumbai, for example, receives only 45 mm in seven months (see graph **C**).

The heavy monsoon rains are very important for growing rice, the staple food in northern India. The rains wash large amounts of soil downhill from the Himalayas. This is deposited on the flood plain of the Ganges to give a deep, fertile soil. Water is also essential for flooding the padi fields in which the rice crop grows. Too much rain can cause severe flooding, too little can mean a crop failure and drought.

C Climate graph for Mumbai

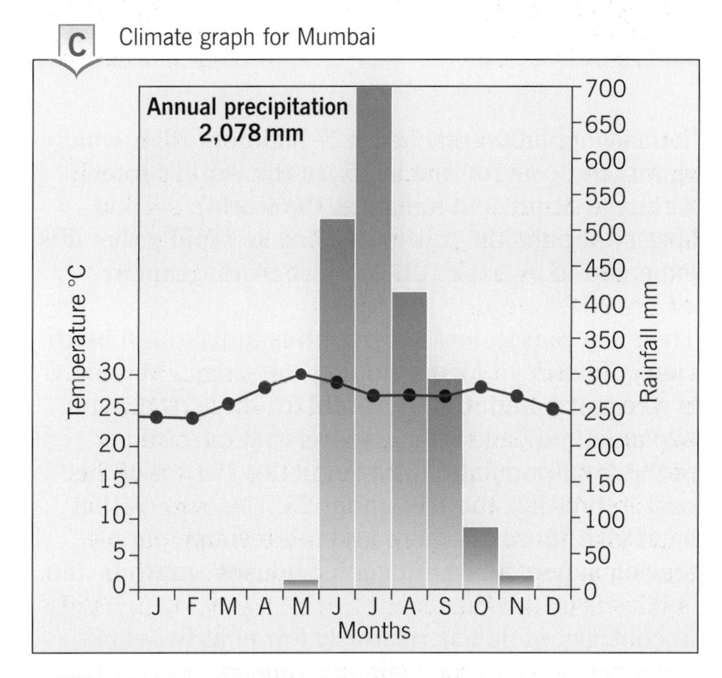

D Flooding in the lower Ganges valley

E

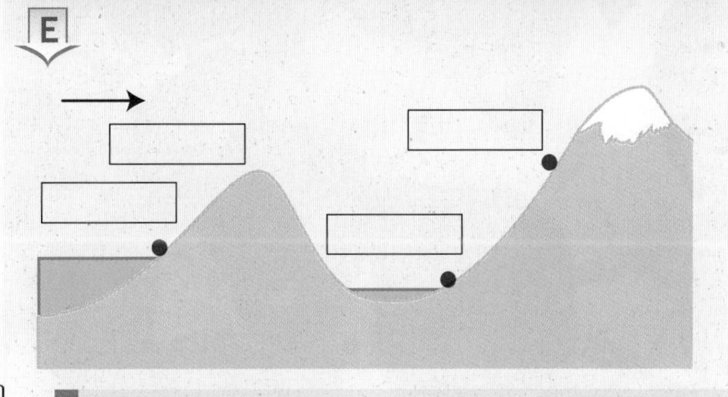

Activities

1 a How did the Himalayas form?
  b Why does northern India experience many earthquakes?

2 Make a large copy of diagram **E**.
  a Label your diagram 'India's south-west monsoon'.
  b Mark on and name:
    the Western Ghats, Himalayas, Mumbai, Delhi, Cherrapunji, River Ganges.
  c Mark on, in the correct places, the following:
    - winds descend to give slightly drier conditions
    - warm, moist winds blow across the Indian Ocean
    - winds forced to rise again giving even heavier rain
    - winds forced to rise, cool and give relief rainfall.

3 a Why is the south-west monsoon important to India?
  b What problems can it create?

Summary

India's relief varies between the flat Ganges Valley and the rugged Himalayas, the world's highest mountain range. Relief rainfall during the south-west monsoon provides water for people and crops.

What are India's main population features?

India's population reached 1.27 billion in 2012, which was 17 per cent, or one in six, of the world's total. Although at present India has the world's second largest population, it is increasing so rapidly that it is expected to overtake China by 2028 (diagram A).

The main reason for this growth is India's high **birth rate**, a feature of all developing countries. Diagram C is a recent **population pyramid** for India (these are explained on page 116). It shows that over 30 per cent of the total population is aged under 15 (it is 18 per cent in the UK) and half under 25. This means that each year there are more and more young people leaving school all wanting jobs, houses and to be fed.

In contrast, India has relatively few elderly people aged over 65 and a low **life expectancy** of 69 years, two other features of a developing country. Yet even in India, people are beginning to live much longer, as they are in almost every country in the world. This growing number of elderly people will increasingly need to be looked after.

A Population of China and India since 1950

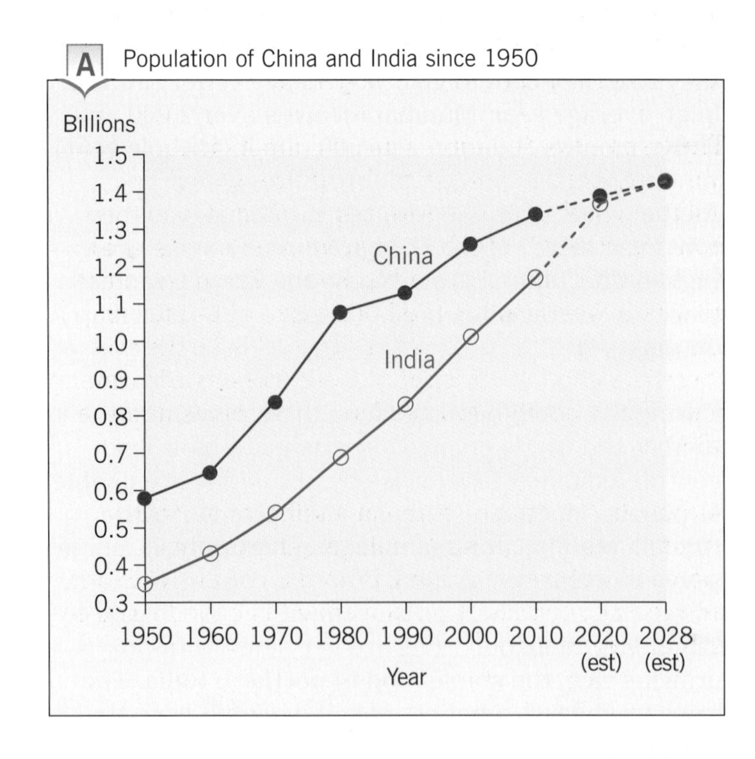

B Children in India

C Population pyramid for India

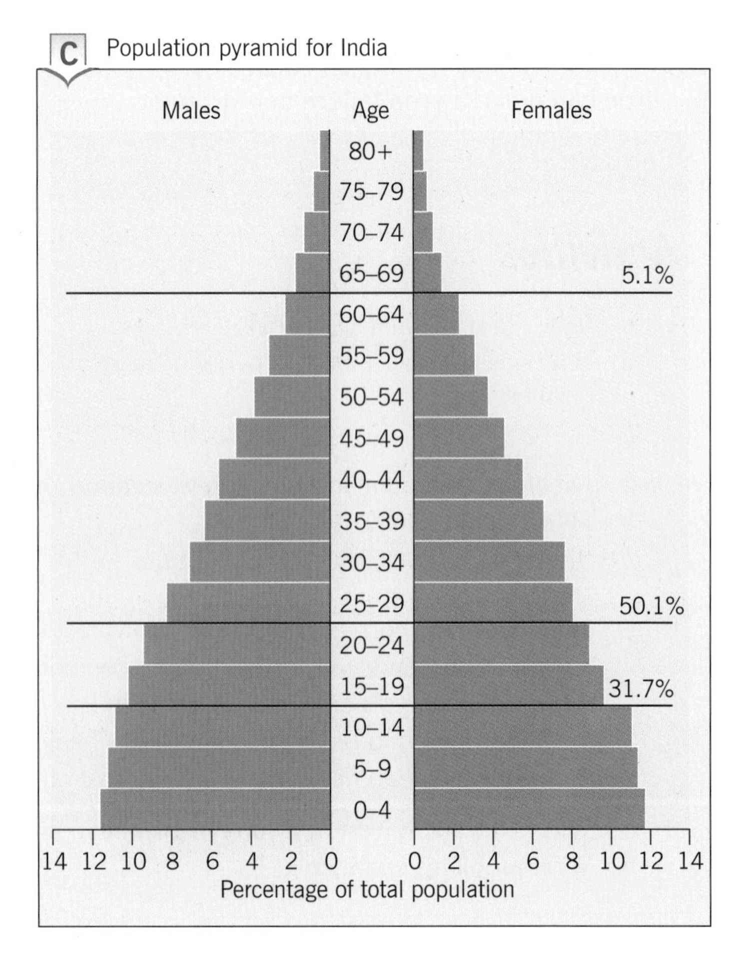

The population of India, as in all countries, is not spread out evenly. Some parts are very crowded while places in the extreme north, where it is mountainous, and the north-west, where there is desert, have very few people living there (see page 54).

The most crowded places are the large cities of Mumbai, Kolkata and Delhi. In all three, over 15 million people live close together, often in poor housing conditions. Their high numbers put a strain on hospitals, schools and transport. The next most crowded area is along the flood plain of the River Ganges where the land is flat, the soil is fertile and there is a good water supply. Here millions of farmers try to grow enough food to feed themselves and their families.

Although 70 per cent of Indians still live in rural areas, often in very large villages, there is an increasing movement away from the countryside and into the large cities. This movement is called **rural-to-urban migration**.

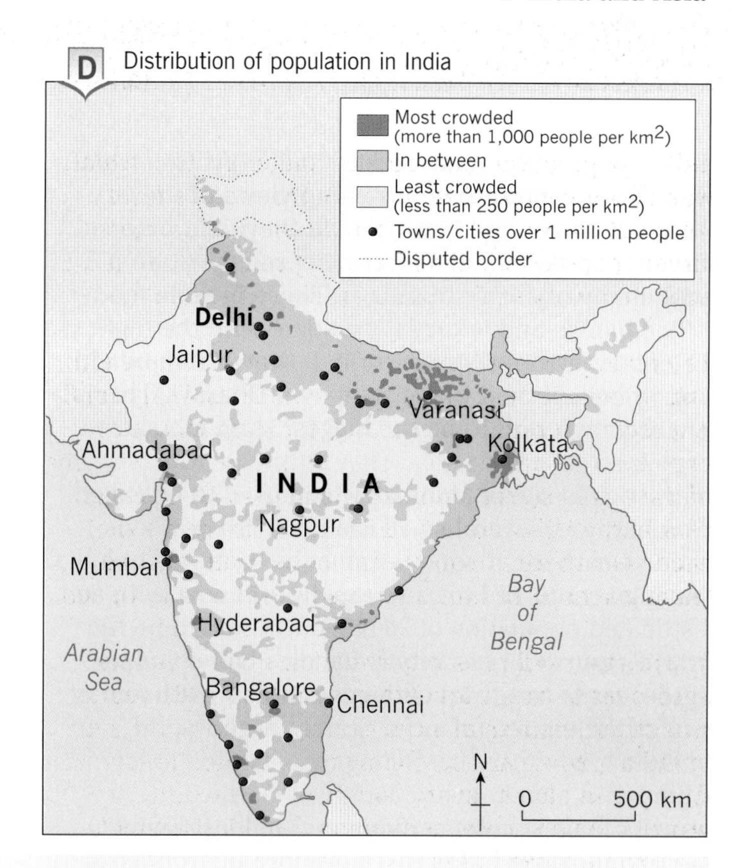

Activities

1 a What were the populations of India and China in:
   i 1950
   ii 2010?
  b Since the 1990s, which country's population has grown the fastest? Give a reason for this.
  c When is it predicted that the two countries will have the same population?

2 a What proportion of India's population is:
   i under 15 years of age
   ii over 65?
  b For each answer, try to give one advantage and one disadvantage of this situation.

3 Look at the statements in **E**. Write out the four that you think are the most accurate.

4 a Why are people leaving the countryside to live in large cities in India?
  b What is this movement called?

5 Describe the problems caused by India's increasing population.

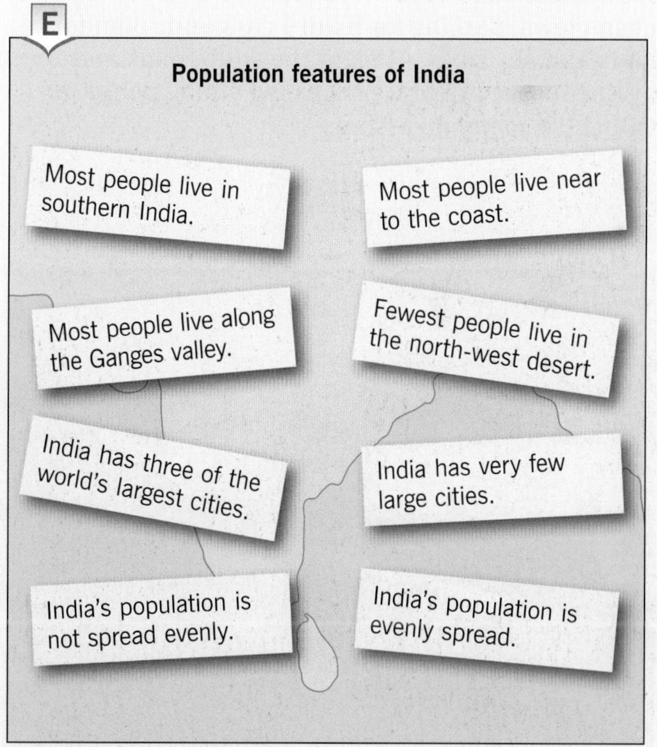

Summary

India will soon have the world's largest population. Many Indians are under 15 years of age. As they grow older most are likely to live in large cities.

What is it like living in Mumbai?

Mumbai is on the west coast of India. Visitors from overseas can get two contrasting views of the city depending on how they arrive. If they land by boat the first thing they see is a prosperous port area with many high-rise buildings. These include modern office blocks and luxury flats, some with the highest prices in the developing world. If they arrive by air, the wheels of their plane skim over Dharavi. This is a shanty town said to be the biggest slum in Asia.

The original site of Mumbai, which used to be called Bombay, was several small islands. With its deep natural harbour, it soon became the 'Gateway to India'. Mumbai continued to grow rapidly and now, with an estimated population of 20 million, is said to be the world's fourth largest city. It handles half of India's trade, earns one-third of the country's wealth and is one of the top ten financial centres in the world.

Mumbai is also home to 'Bollywood', the world's biggest cinema industry. The best living accommodation offers three or more bedrooms, ocean views, parking for luxury cars and room for maids and servants. At night the wide roads and pavements are full of well-dressed young people visiting the many nightspots.

A Mumbai

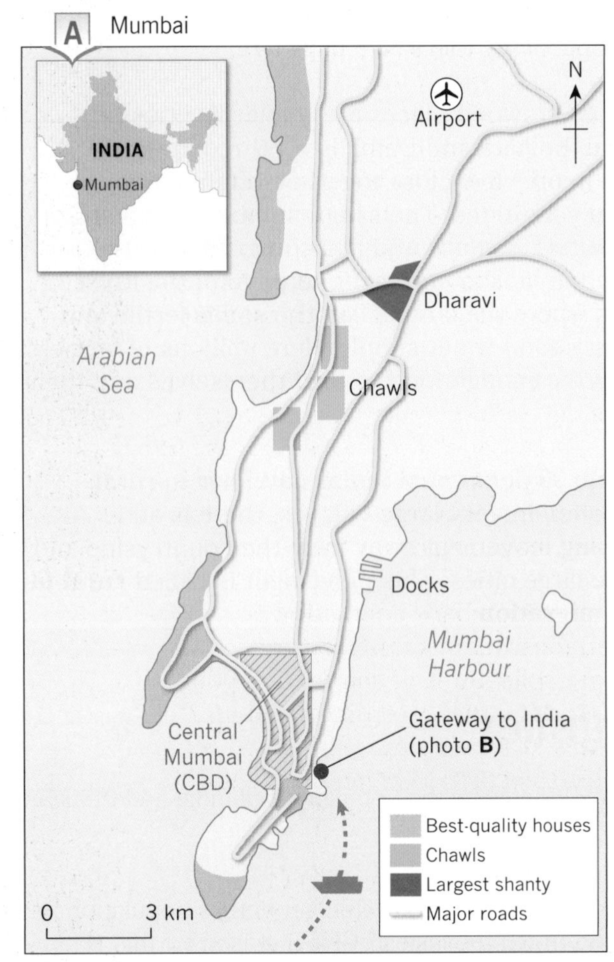

B Port of Mumbai and the 'Gateway to India'

Almost unique to Mumbai is the **chawl**. These are buildings with four or five storeys. They were built in the late 19th century for the thousands of workers in Mumbai's textile mills. Chawl means 'corridor'. Each storey has one long corridor, off which there are up to 20 individual tenements (dwellings). A tenement usually has two rooms and a balcony. One room is used for living and sleeping, the other for cooking and eating. The balcony overlooks the street and helps to create a friendly, communal atmosphere.

Families on each corridor have to share a common lavatory block but have access to electricity. Adults are likely to have a full-time job, even if it is relatively low skilled and poorly paid, to pay for their rent.

Over half of Mumbai's population is crowded together in **shanty settlements**. These are collections of shacks and poor-quality housing that are common in most large cities in developing countries. They often lack electricity, a clean water supply and any system of **sewage** dispersal.

The largest shanty in Mumbai is Dharavi. Over one million people live here in houses made from bits of wood, cardboard, plastic and corrugated iron. The area is a maze of narrow lanes and is overcrowded and polluted. Infectious diseases such as dysentery and hepatitis are common. Few people have jobs and most live by recycling waste materials such as plastic, tin cans and old oil drums.

C A chawl in Mumbai

D Dharavi

Activities

1. a Look at map **A**. Describe the location of the best-quality housing in Mumbai.
   b Why do you think many 'Bollywood' stars live here?
2. a What is a chawl?
   b List three good points about living in a chawl.
   c List three bad points about living here.
3. a What is a shanty settlement?
   b With the help of photo **D**, list four problems of living in a shanty settlement such as Dharavi.
   c How might people living in a shanty settlement like Dharavi earn a living?

Summary

There are two sides to cities in developing countries. In Mumbai the more wealthy people live near to the city centre and the least well-off live further away in shanty settlements.

What is it like living in a village in India?

In India today more people still live in villages in rural areas than in large cities. Most of them are farmers.

The village of Ooruttukula is in the state of Kerala in the south of India. Kerala is one of the more wealthy Indian states. This is partly because its climate and soil are suited to agriculture. It is also partly due to the fact that its state government encourages education, so people are better informed about use of the land.

Most Indian villages are strung out along the one main road. The road itself is likely to be dusty during the dry season of the north-east monsoon, and flooded during the wet monsoon. It is often busy with cars, cyclists, handcarts, trishaws and people.

The road is lined on either side with shops. These sell locally grown food and other goods. The shops and houses are built with mud bricks. The roofs may be of palm fronds woven together, or tiles. Many shops and houses have a veranda to give protection from the hot sun and shelter from the heavy rain.

A Road through a village

B Village shops

C Market trader, Trivandrum, Kerala

D Preparing tapioca

Until recently water had to be taken from a local river or stream. Now, thanks to the help of international charity organisations such as WaterAid (page 97), the village has its own water pump. This has three main advantages:

- It provides a reliable source of clean water.
- It has helped improve the health of the village.
- People no longer have to walk to the river to get their drinking water.

However, villagers continue to use the river for washing, and the disposal of sewage is still a problem.

For fuel, people use local wood, palm fronds and crop waste. Some people use animal dung but this means it cannot be used on their fields as a fertiliser.

The farmers walk the short distance to their fields. The main crop is rice. The climate is hot and wet, and the soils are rich, so two crops can be grown each year. Bananas, ginger, tapioca and vegetables are also grown. Tapioca, or cassava, is a staple food for the Hindu community in Kerala. Once grown, its roots are skinned and boiled to make flour (photo **D**)

The cow is a valuable animal:

- It is used to plough the fields and thresh the grain.
- It transports people and their goods.
- It provides milk.
- It provides dung: mixed with straw it can be used as a fuel, or mixed with mud it can be used as a building material. It can also be used as fertiliser.

E Village pump

F Washing and drying clothes by the river

Activities

1 Rewrite the following paragraph choosing the right word from the pair in each bracket.

> Kerala is in the (north/south) of India. It has a (desert/monsoon) climate with (high/low) temperatures all year and a very (dry/wet) summer. It is mainly a (rural/urban) area and the state is relatively (poor/wealthy).

2 Using table **G** as a guide, describe the differences between a village in Kerala and one in your local area.

G

	Village in Kerala	Village in your local area
Road		
Transport		
Building materials		
Shops		
Fruit/vegetables		
Farming methods		
Crops/animals		
Water supply		
Sewage		
Fuel		

Summary

Half the population of Kerala still live in villages where the traditional way of life is only slowly changing.

How interdependent is India?

Most countries would like to be **independent**. They want to make their own decisions on how they develop and how their people live. However, no country has everything it needs. It may lack certain foodstuffs, minerals, energy resources, skills or technology. If it wants these things, it will have to work with other countries. Only then can it develop and improve its own standard of living. When countries work together and rely upon each other for help, they are said to be **interdependent**.

One of the main ways that countries become interdependent is by selling goods to each other. They buy things they need or would like to have. They then sell things to pay for what they have bought. This exchange of goods and materials is called **trade**.

Goods sold to other countries are called **exports**. Goods that are bought from other countries are known as **imports**. The difference between the costs of imports and exports is the **trade balance**.

Unfortunately, in the past it has been the richer, developed countries that make most money from trade. They are said to have a **trade surplus**. Meanwhile the poorer, developing countries earn far less and have a **trade deficit**.

Diagram **B** shows India's balance of trade. It is typical of a developing country. This is because the cost of its imports is greater than money earned from its exports. However, since 1990, the trade of developing countries such as China and India has increased at a faster rate than the trade of developed countries. India's trade deficit is getting less as its economy grows.

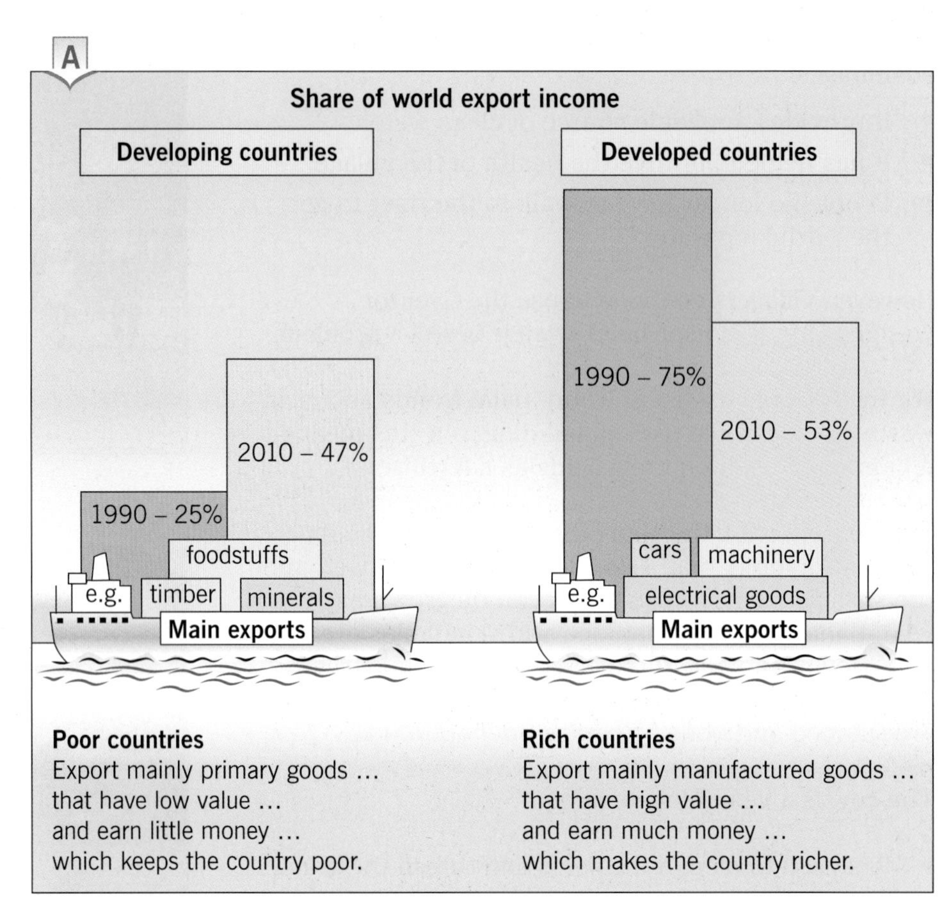

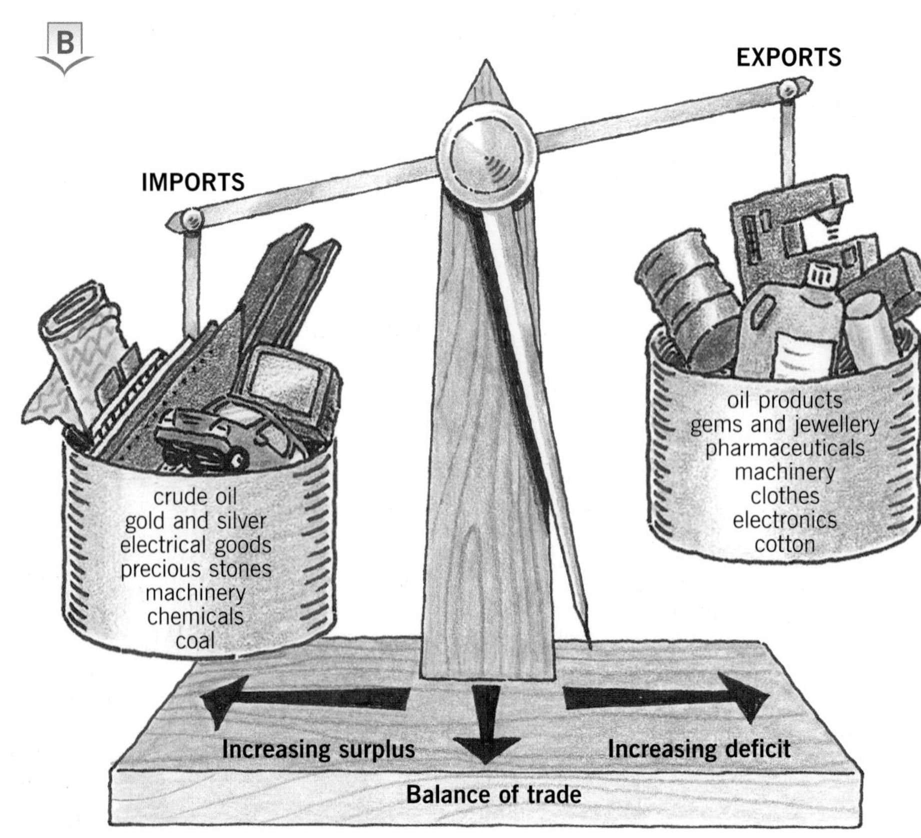

Another important question is: which countries does India trade with? When it became independent in the 1940s, nearly all of its trade was with the UK and Europe. However, this was to change so that even by 1990 most trade was with Asia (diagram **E**). The rapid growth of India's economy since 2006 has seen a three-fold increase in both imports, mainly oil from the Middle East and manufactured goods from China, and also exports, largely to Asia (diagram **D**).

C Container ship in Mumbai docks

D India's imports and exports 2012

Imports $235bn

Region	%
Middle East	26%
Europe	19%
North America	14%
China	12%
Rest of Asia	11%
Rest of world	18%

Exports $142bn

Region	%
Rest of Asia	26%
Middle East	21%
Europe	19%
North America	12%
China	5%
Rest of world	17%

Activities

1 a What does it mean when a country is said to be 'interdependent'?
  b Give three examples of India's interdependence with other countries.

2 Give the meaning of each of the following terms. You may need to use the Glossary on pages 118–120.

- trade
- imports
- exports
- trade surplus
- trade deficit

3 a Why do most developed countries have a trade surplus?
  b Why do some developing countries have a trade deficit?

4 a Look at diagram **E**. How have the regions that India trades with changed in the last 20 years?
  b Which places do you think India will do most trading with in 20 years' time?

E Direction of India's trade

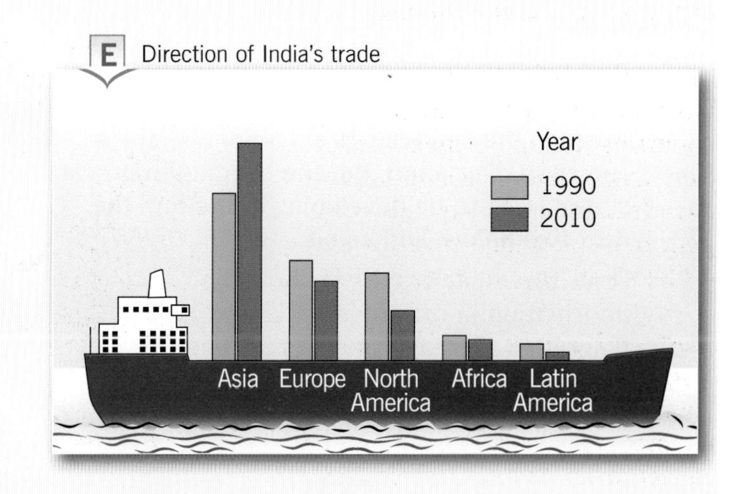

Summary

As India's trade is growing rapidly, it is becoming increasingly interdependent with other countries, especially those in Asia.

The India enquiry

All countries are different. Some are rich and have a high **standard of living** whilst others are poor and have a lower standard of living. Countries that differ in this way are said to be at different stages of **development**.

Japan and the UK are examples of rich countries that are said to be **developed**. Kenya and India are at the other end of the scale. They are mostly very poor and are said to be **developing**.

Your task in this enquiry is to make a report on India's level of development. It is best if you can work with a partner so that you can share ideas and discuss different points of view. You should present your report in three parts and use writing, maps, graphs and diagrams where appropriate.

A

How developed is India?

1 Introduction

- **a** What is meant by the terms 'development' and 'standard of living'?
- **b** Explain how the three factors in diagram **A** help to measure development.

2 Main part

- **a** Give each of the nine drawings 1 to 9 in diagram **C** a heading.
- **b** Arrange the headings in a diamond shape as shown in diagram **B**. Put the heading that best shows India to be developing at the top, the next two below, and so on.
- **c** Draw three bar graphs to show the information in drawings 10, 11 and 12 in diagram **C**. Arrange the bars as follows:
  - wealth – highest income on the left
  - education – most able to read and write on the left
  - food – highest-quality food supply on the left.

3 Conclusion

- **a** In terms of wealth and other economic factors, what evidence is there to suggest that India is still a developing country?
- **b** How developed is India in terms of social and cultural factors?
- **c** India is said to have one of the world's fastest-growing economies. How is this helping the country to develop and improve its standard of living?

B

India – measures of development

C

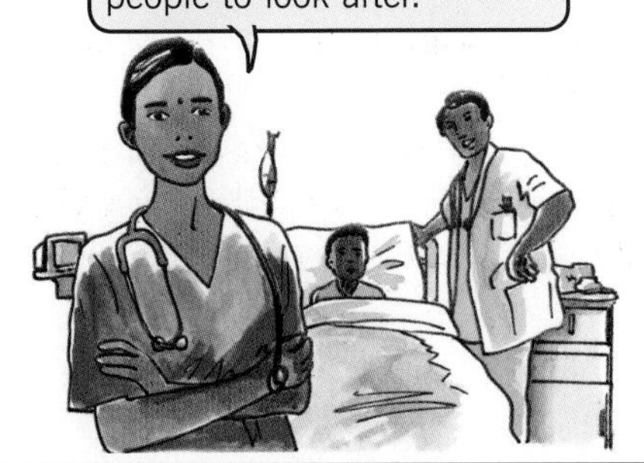

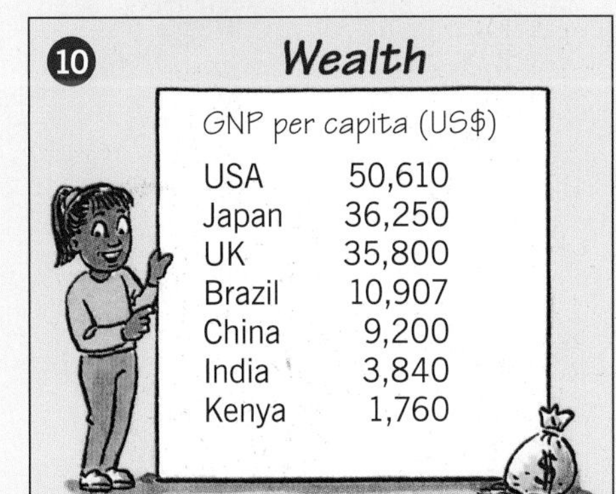

10 **Wealth**

GNP per capita (US$)	
USA	50,610
Japan	36,250
UK	35,800
Brazil	10,907
China	9,200
India	3,840
Kenya	1,760

11 **Education**

% adults able to read and write	
Japan	99
UK	99
USA	99
China	92
Brazil	89
Kenya	87
India	74

12 **Food**

Daily calorie supply	
USA	3,770
UK	3,450
Japan	3,120
Brazil	3,010
China	2,970
India	2,473
Kenya	1,860

What are world issues?

What is this unit about?

This unit looks at some of the major problems facing our world. These problems are called world issues because they affect the lives of people across the world. Such issues may only be solved by international co-operation.

In this unit you will learn about:

- the causes and effects of global warming
- how our use of energy may change
- the problems of food supply and water shortages
- the poverty problem.

A What London may look like after a major flood

Why is it important to know about world issues?

One very good reason is that some issues are likely to affect you personally. For example, global warming will affect the way you live, having too much food will affect your health, and requests for aid will give you the choice of helping others.

But there is much more to it than that. Learning about the problems facing our world can help you appreciate the need to look after our planet. It can help you become a global citizen, interested in the state of the world, aware of the problems and willing to do your bit to solve them.

- Look at photos **A** and **B**. For each one:
  - describe what has happened
  - describe the problems
  - suggest what caused the problems
  - suggest what might be done to reduce the problems.
- How would you feel about living in place **B**?
- List the problems that the family in photo **C** may have.

C The slums of Delhi

B Children playing on the dry riverbed of the Jialing river, China

What is climate change?

The earth's climate has changed many times. Sometimes it has been much warmer than it is now, and some times much colder. For example, in a colder period that ended just 11,000 years ago, the UK was covered in snow and ice. This is called an **ice age**. Graph **A** shows changes over the last 1,000 years. The term **climate change** is used to describe these changes.

As the graph also shows, the world is now much warmer than it has been in the past. In the last century, average temperatures rose by almost 1°C, with the greatest increases in the last 40 years. This century has already seen the warmest year ever recorded. The term **global warming** is used to describe the heating up of our planet.

Climate change is thought to be due to the **greenhouse effect**. As diagram **C** shows, the earth is surrounded by a layer of gases including carbon dioxide. These keep the earth warm by preventing the escape of heat that would normally be lost to the atmosphere. The gases act rather like the glass in a greenhouse. They keep the heat in but prevent most of it from getting out (diagram **B**).

The burning of fossil fuels such as coal, oil and natural gas produces large amounts of carbon dioxide which, as diagram **D** shows, is the main greenhouse gas. As the amount of this gas increases, the earth becomes warmer.

A Global temperatures and carbon emissions

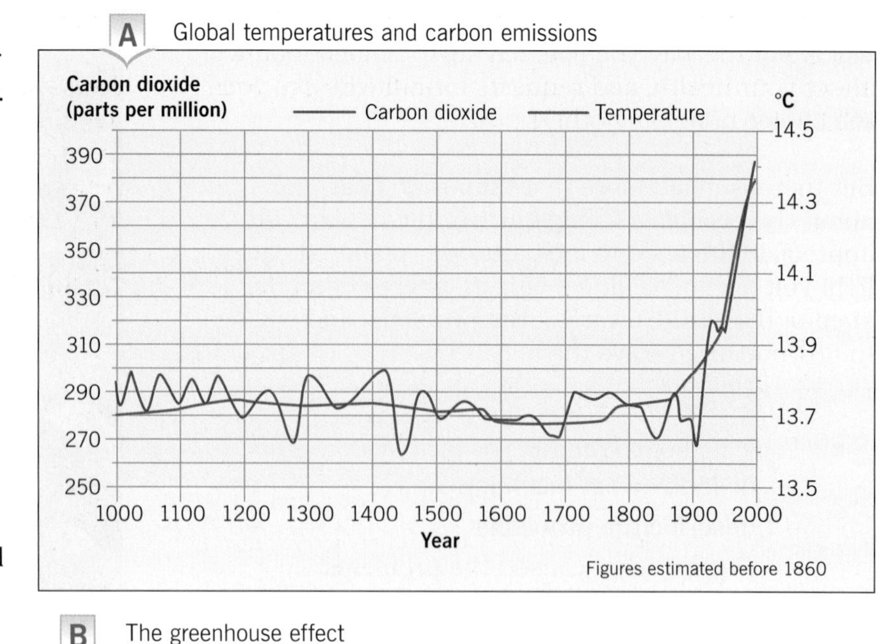

B The greenhouse effect

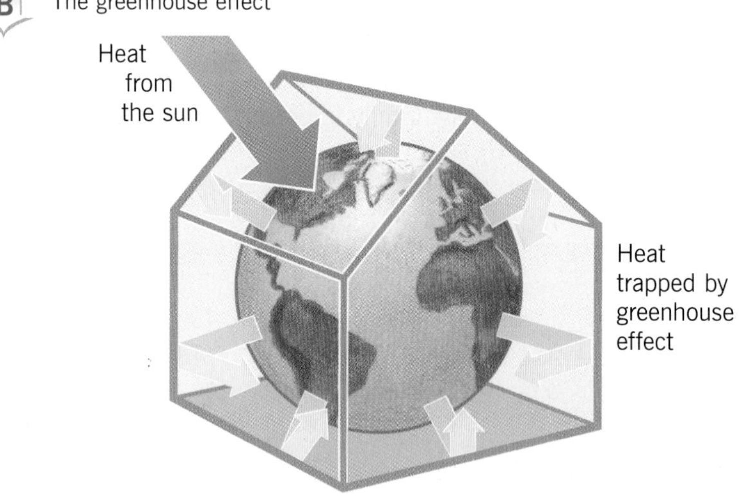

C Causes of the greenhouse effect

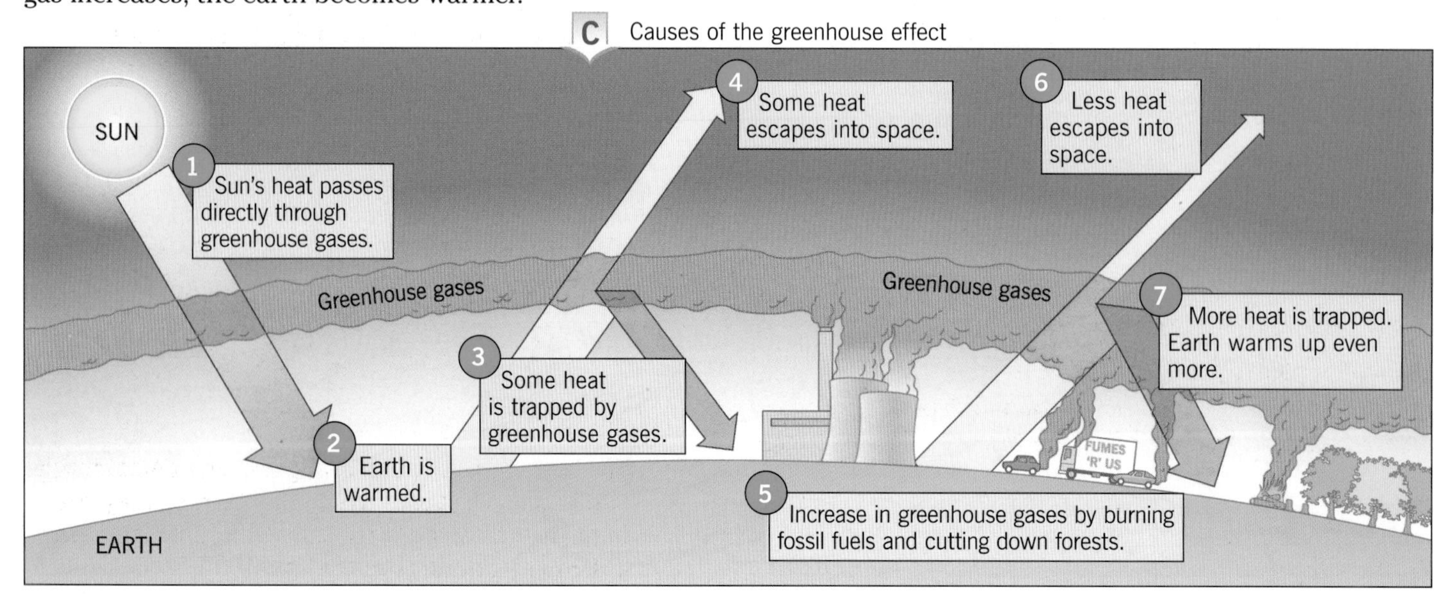

Most scientists believe that over the next hundred years temperatures will rise between 2°C and 4°C. Some think it may be even more than that but a few believe that other effects might cancel out this warming. Whoever is right, it is fairly certain that in the years ahead it is unlikely that conditions on earth will be the same as they are now.

Global warming is a world problem. Almost every country contributes in some way to producing greenhouse gases. Global warming affects every corner of the planet.

International agreement is needed if we are to reduce greenhouse gases and slow down global warming. So far, this agreement has been difficult to achieve and global warming continues to be a real problem for our world.

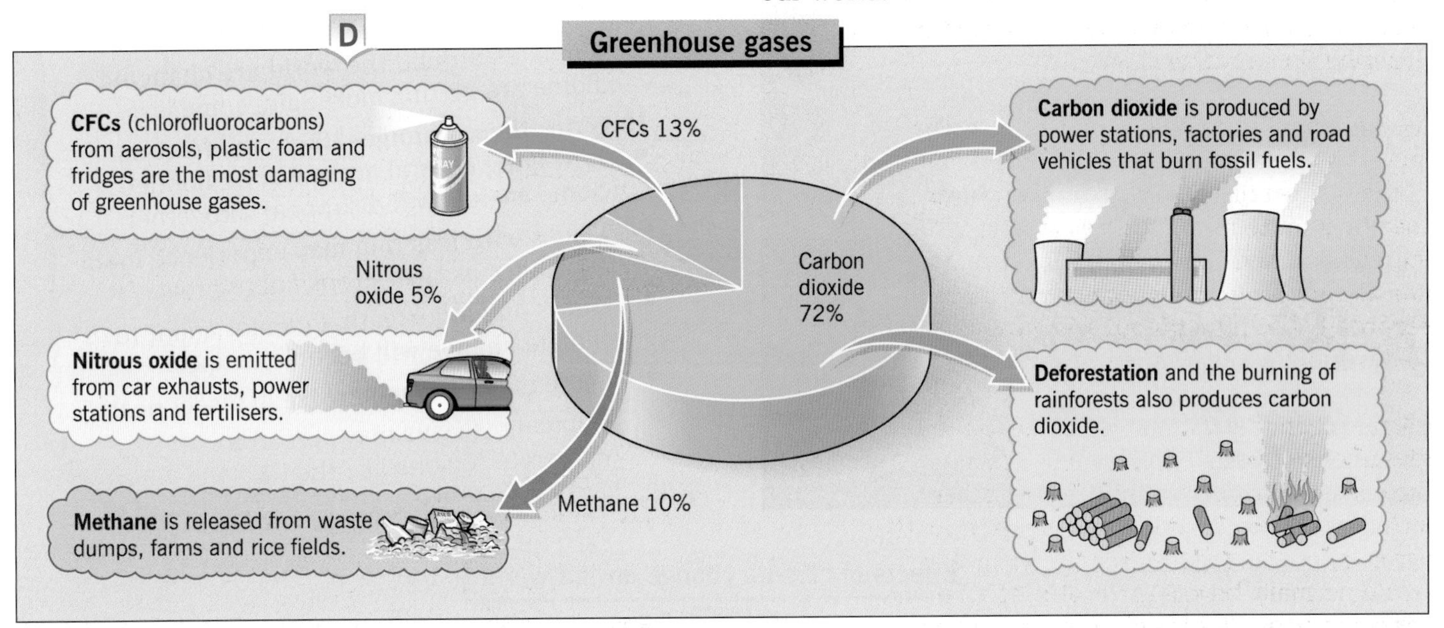

Activities

1 Look at graph **A**.
   - a Which were the three coldest years? What was the temperature in each of those years?
   - b Which was the hottest year? What was the temperature in that year?
   - c Describe the change in global temperatures.
   - d Describe the change in carbon dioxide levels.
   - e Explain the link between the two.

2
   - a What is climate change?
   - b What is global warming?
   - c What is the greenhouse effect?
   - d Give two reasons why global warming can be described as a world problem.

3
   - a Make a larger copy of diagram **E**.
   - b Add labels to your diagram to explain how the burning of fossil fuels may cause climate change.

4
   - a Which greenhouse gases result from burning coal, oil or natural gas?
   - b Why is international agreement needed to reduce greenhouse gases?
   - c Make a list of ways in which you may have contributed to global warming in the last week.

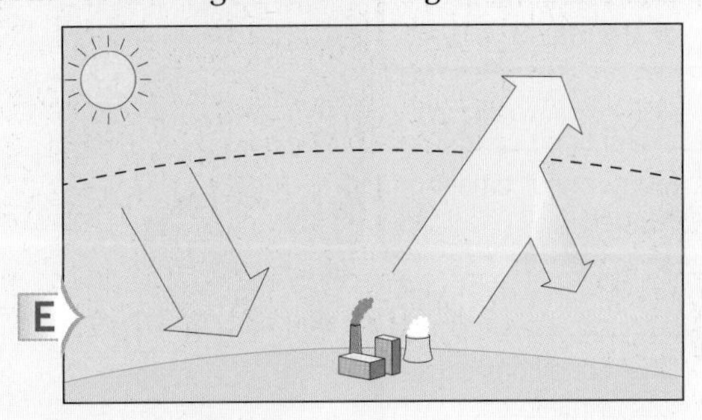

Summary

The earth's climate has changed many times mainly due to natural causes. Since the 19th century the burning of fossil fuels has caused a rapid increase in temperatures around the world.

What are the effects of climate change?

Nobody knows exactly what the effects of climate change will be. Some of the effects will no doubt be harmful, but others may bring benefits. Some of the effects predicted by scientists are shown on these two pages.

A Penguins on melting sea ice, Antarctica

B

How is climate change affecting our world?

- Sea temperatures are rising, causing the water to expand and the sea level to rise.
- Ice caps and glaciers are melting, causing sea level to rise even further.
- Climates around the world are changing. Some are getting more rain, some less.

How are these changes likely to affect us?

- Low-lying coastal areas will be flooded. Some islands will disappear altogether.
- Places with less rain may experience food shortages as their crops fail to grow.
- Plants and animals that cannot adapt to climate change will become extinct.
- There might be an increase in insect pests.
- Tropical diseases may spread to temperate regions like the UK.

C Effects of climate change on the world

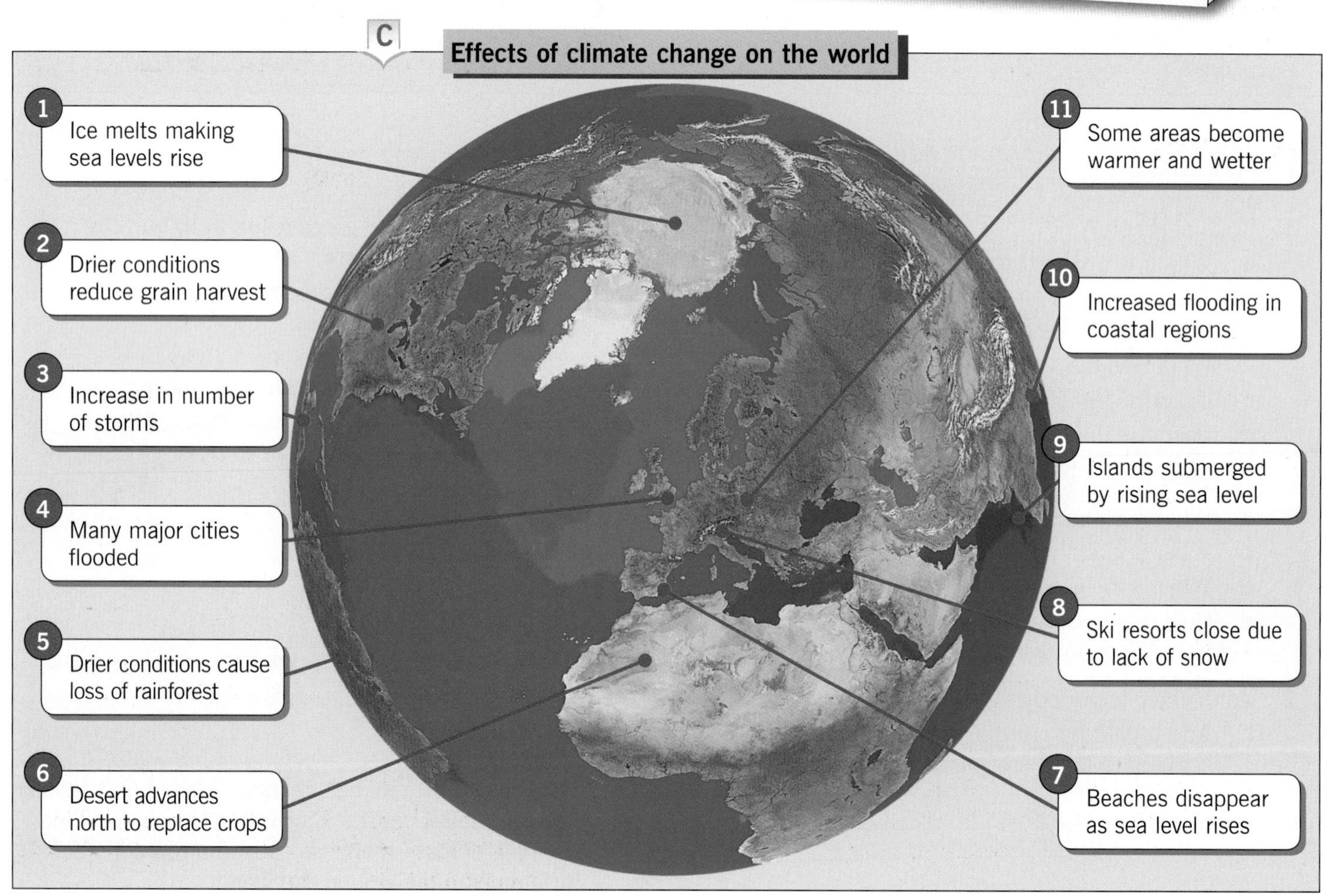

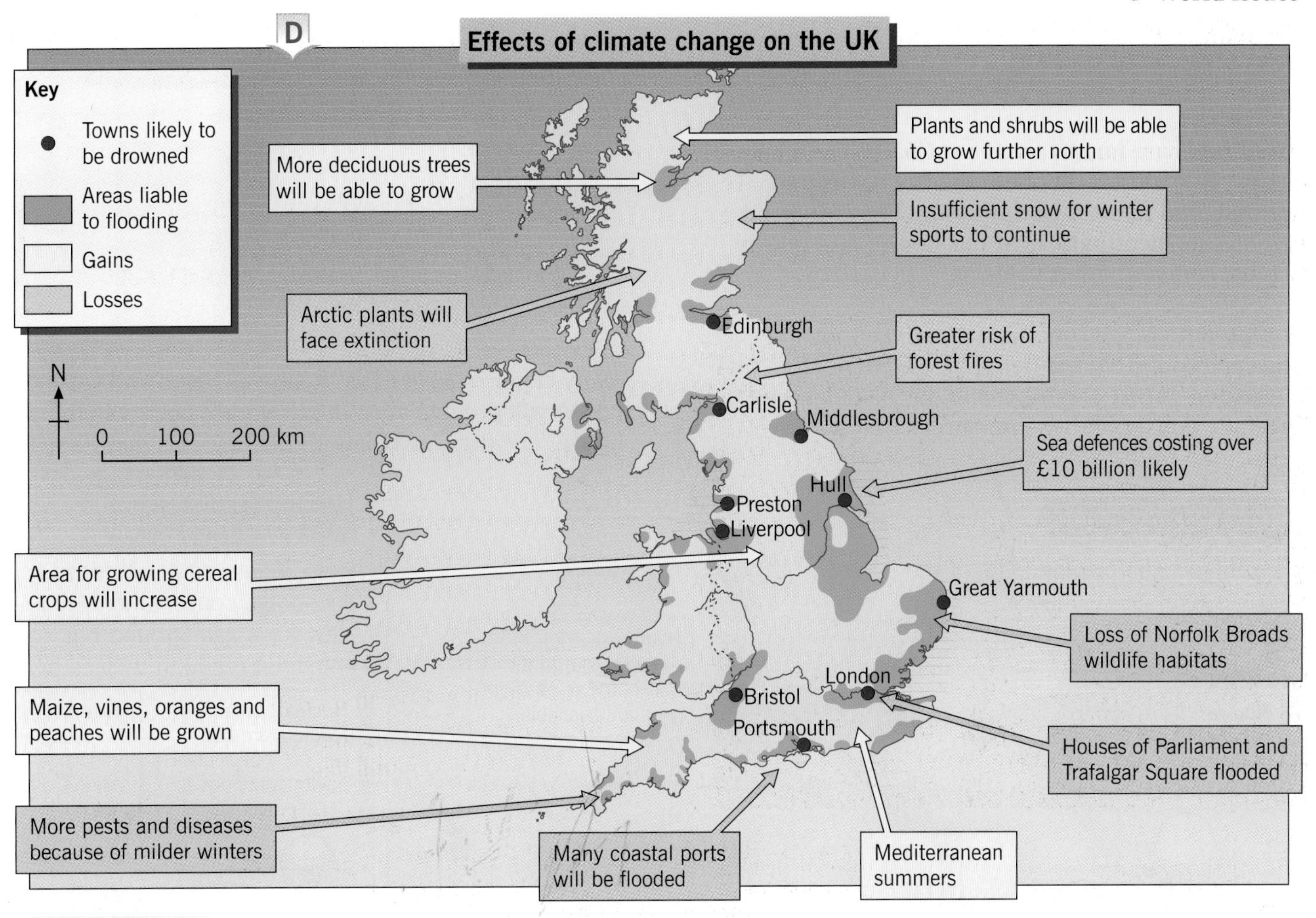

Activities

1 Look at satellite photo **C**. Match the locations on the photo with the places below. The world map on the back cover will help you.

- Caribbean Sea • Sahara Desert • Arctic Ocean
- Amazon • Mediterranean Sea • Bangladesh
- Maldives • Europe • Great Plains • Alps • London

2 a Make a larger copy of diagram **E**.
b Put the statements in the panel below into the correct boxes coloured yellow.
c Complete the 'Effects' boxes (coloured green) with six examples from box **B** or photo **C**.

- Water expands • Sea level rises • Ice cap melts
- Plants and wildlife affected • Climate changes

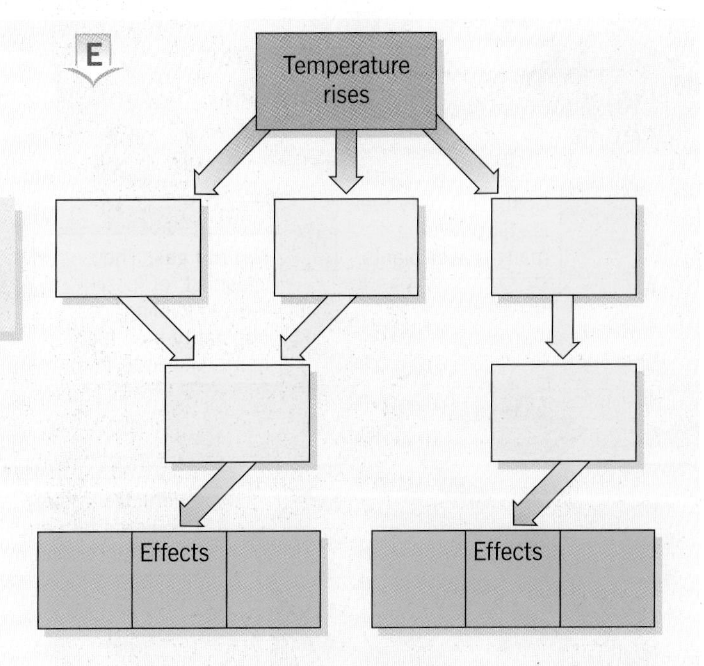

3 Describe how climate change in the UK:
a will bring benefits to farmers
b will cause most problems for people
c will affect the area where you live.

Summary

Climate change will cause serious problems throughout the world. Some of the effects may bring a number of benefits.

How can our energy use change?

Fossil fuels are important to us. Coal, oil and natural gas provide almost all of our energy needs and are essential to modern-day living. They are used to generate electricity, heat our homes and power planes, cars and other vehicles.

But fossil fuels are running out and, as they are non-renewable, cannot be replaced. We need to find alternative forms of energy and, in the meantime, carefully conserve those resources that we have.

When will we run out?
At our present rate of use ...
- Oil ... in about 50 years
- Gas ... in about 60 years
- Coal ... in about 250 years

Many experts believe that hydrogen, wind and solar power will provide most of the world's energy in the future. The drawing below shows how this may happen.

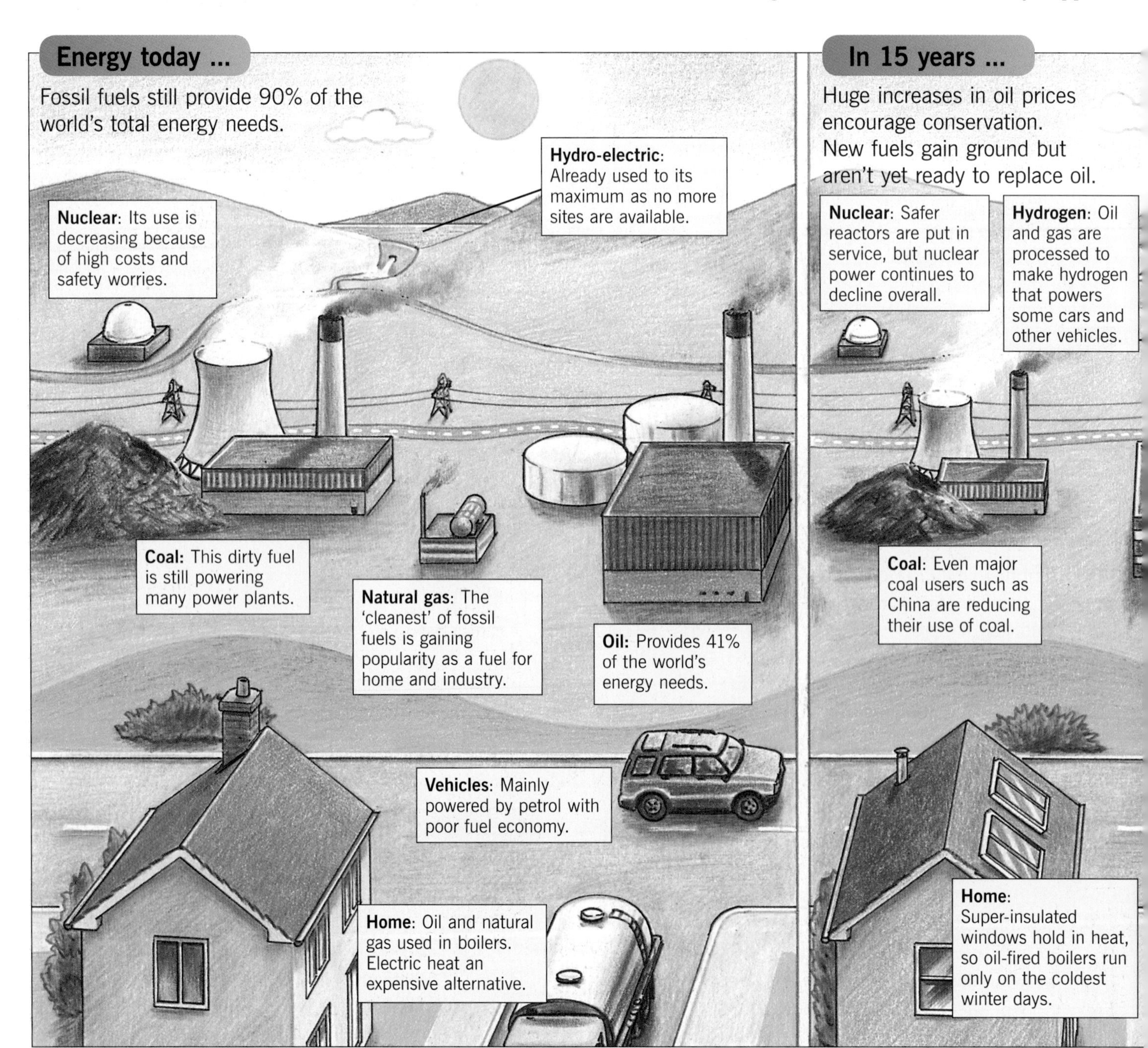

Activities

1. a Which energy resource is most used today? Give a list of uses.
   b Which energy resource is likely to be most used in 50 years' time? Give a list of uses.
2. During the next 50 years, which energy sources are expected to be:
   a used less b used more?
3. How will changes in energy use in the next 50 years affect global warming? Give reasons for your answers.
4. How would it affect you if oil ran out? Think about your day-to-day life.
5. In what ways do you think the energy sources that may be used in 50 years' time are better than the ones we use today?

Summary

There are limited supplies of fossil fuels and, once they are gone, they cannot be replaced. Other forms of energy need to be developed in their place.

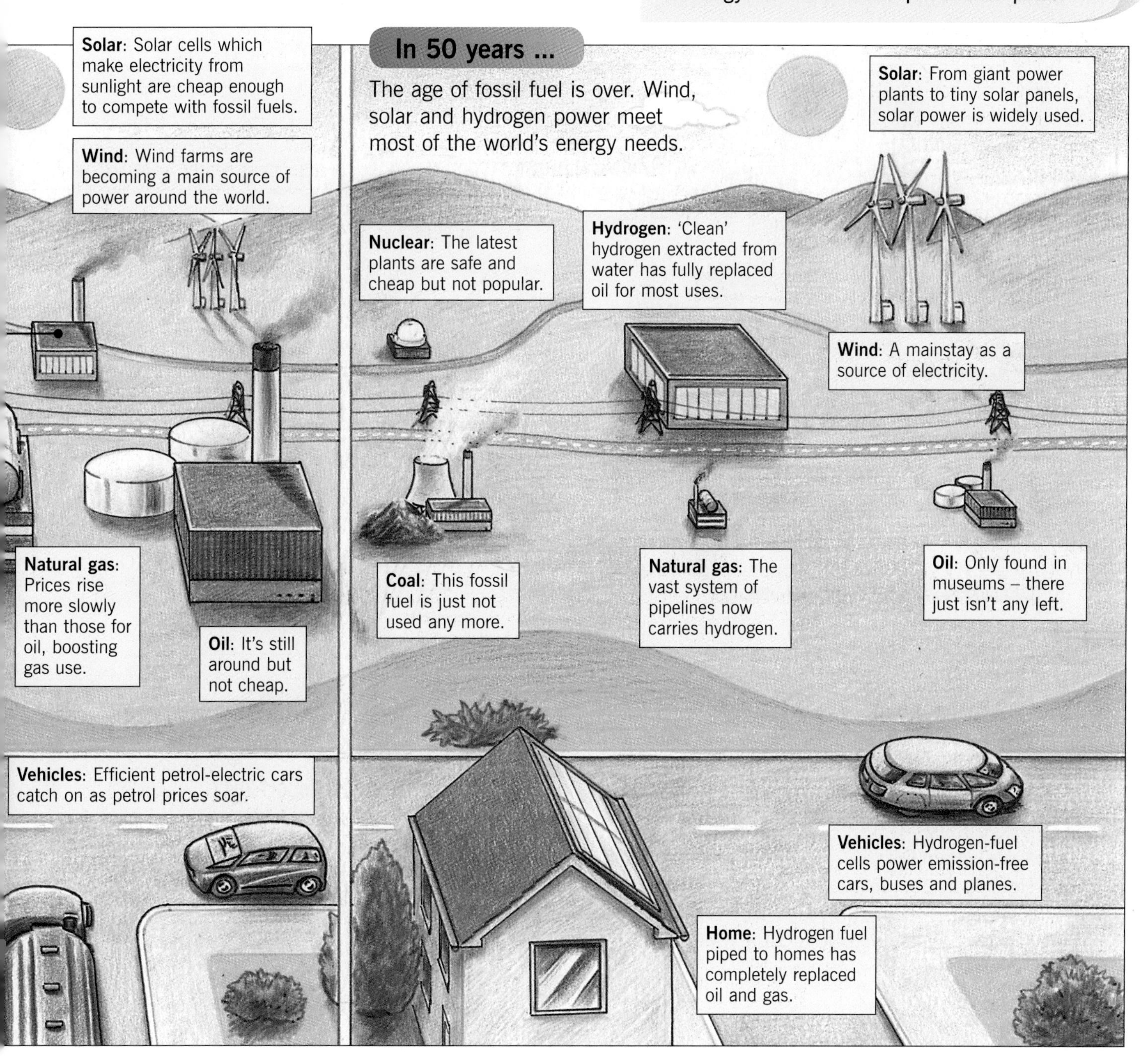

What is the water problem?

In Britain we take it for granted that, apart from the occasional drought, a reliable supply of clean water is always there when we need it. This is because:

- we get rain spread fairly evenly throughout the year
- we have the money and technology to create reservoirs from which water can be piped to our homes and places of work and leisure.

This is not the case for many people in other parts of the world, especially those living in developing countries. The United Nations claims that two out of every five of the earth's population lack a safe and reliable supply. It also suggests that there is enough fresh water to support five times the present world's population. However, this water is either unevenly spread or is hard to reach.

As the demand for water grows, this normally renewable resource is increasingly under pressure. Diagram **A** gives some reasons why developing countries may be short of water. Diagram **D** shows some of the effects of water shortages.

A Causes of water shortage

Activities

1 Look at diagrams **A** and **B**.
   - **a** Give reasons why so many people are short of water.
   - **b** Why are water shortages greatest in developing countries?

2 Look at diagram **D**.
   - **a** How many people are likely to be short of water in 2050?
   - **b** Why will it be difficult to provide water for so many people?

B

UNITED  NATIONS

UN Report 2012

- 783 million people around the world are without clean drinking water.
- In the poorest countries of Africa, over 40% of the population are without clean water.
- Water-related diseases claim over 2.6 million lives a year – that is one death every 12 seconds.

The United Nations predicts that by 2050 the number of people who will be short of fresh water will rise to over three in every five. It also predicts that water shortages may cause problems between countries that, at present, rely upon the same water supply.

There are no easy solutions to the world's water shortage. Building dams to create reservoirs is out of favour. Many people in the poorest of developing countries have to rely upon help from voluntary organisations like WaterAid (diagram **C**). WaterAid collects money through donations in the UK. It then uses this money either to provide clean water or to improve sanitation in poor countries.

C

WaterAid

- The UK's only charity dedicated solely to providing safe, clean water, sanitation and hygiene education to the world's poorest people. It believes these basic services to be essential to life. Without them vulnerable communities have little chance of escaping the stranglehold of disease and poverty.
- So far WaterAid has helped 10.2 million people. Its aim is to help beat the threat of death and misery caused by a lack of clean water and give people hope for a better future.

D Effects of water shortage

3 Look at the statistics on the back cover.

- **a** List the countries in order of:
  - wealth
  - access to clean water.
- **b** Write a sentence to describe the link between the wealth of a country and its access to clean water. Suggest reasons for this.

4
- **a** Photo **E** shows clean water provided by WaterAid in Nepal. How can organisations like WaterAid help in providing clean water to people living in developing countries?
- **b** How might **you** be able to help?

E

Summary

Nearly half of the world's people have no reliable supply of clean water. The situation is expected to get worse.

Food – too little or too much?

Many of us living in a developed country such as the UK are used to having at least three good meals a day. Added to that are various snacks that we take any time we are hungry, thirsty or, in some cases, just bored. In contrast, many people who live in poorer, developing countries consider themselves lucky if they get one good meal a day.

For people to have a satisfactory diet they need:

- the correct **quantity** of food – the amount of food a person eats is measured in calories
- the correct type or **quality** of food – a healthy diet consists of proteins (meat, eggs and milk), carbohydrates (cereals and potatoes) and vitamins (fruit and vegetables, meat and fish).

Diet can be a problem in both poor and rich countries.

The United Nations says that there is enough food produced each year to feed everybody in the world. Unfortunately while rich countries like the USA and those in Western Europe produce more than they need, many poorer countries, especially in Africa, do not produce enough.

A

The United Nations

- The UN claims that 1,200 million people were short of food in 2012.
- It predicts that this number will have risen to 4,200 million by 2050.

Some causes of food shortages

B

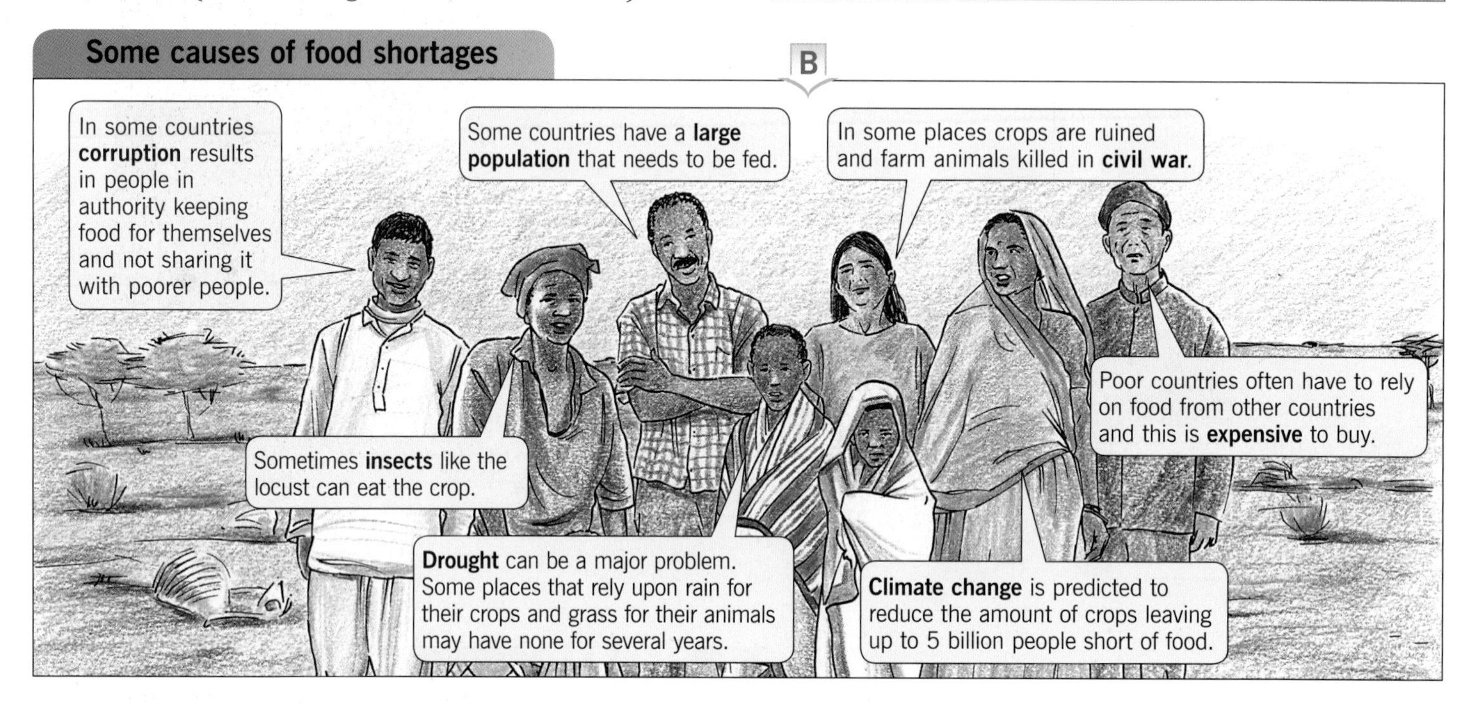

Poor countries

By 2050 half of the world's population is likely to be underfed (diagram **A**). Diagram **B** gives several reasons why certain places are short of food. A person who does not get enough to eat or the right type of food is likely to suffer from **malnutrition**. Malnutrition usually results from poverty when people have not got enough money to buy food, rather than because there is too little food for the number of people. Diagram **C** shows some of the effects of malnutrition.

Rich countries

A person who either eats too much or who eats the wrong type of food is also unhealthy. An increasing number of people in developed countries like the USA and the UK are overweight. They are said to be **obese**. Diagram **C** also shows some of the effects of being overweight. In the UK:

- what we like to eat is not always good for us
- what we should eat we often do not like.

Your school may be one of those that has been encouraged by the television and the government to change its dinner menus to provide a healthier diet.

C Calorie intake

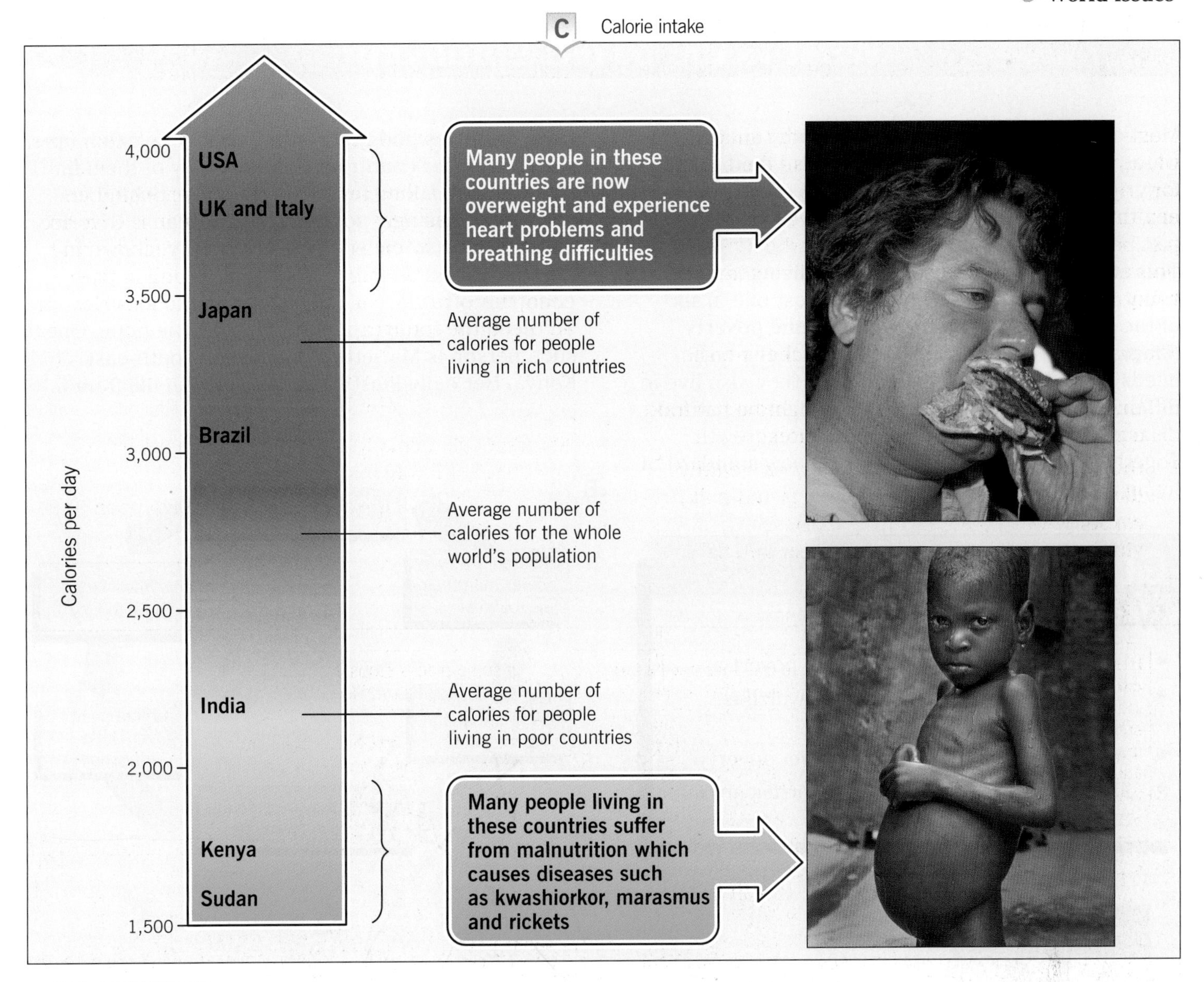

Activities

1 Using diagram **B**, list the causes of food shortages under these headings:

Natural causes | Caused by rich countries | Caused by poor countries

2 a Describe what would be a good diet for yourself. Include the number of calories and the type of food that would give you a healthy diet.

b What changes to your present diet would help you and your family to live a healthier life?

D

	% obese	% malnourished
1		
2		
3		

3 a Look at the statistics on the back cover. On a copy of table **D**, list in rank order, with the highest first, the percentage of people in the six countries suffering from

- obesity
- malnutrition.

b What is obesity?

c What are the causes and effects of obesity?

d What is malnutrition?

e What are the causes and effects of malnutrition?

Summary

There are considerable differences in the quantity and quality of food supplies between rich and poor countries. A poor diet can cause many illnesses.

What is the poverty problem?

Most of us in the UK live in houses with running water, sewerage and electricity. We also tend to take for granted other basic needs such as education, health care, jobs and access to plenty of food.

This is not the case for many people living in poor countries. It is now believed that almost one in six of the world's population live in **extreme poverty** (diagram A). These people not only lack the basic needs available in rich countries, but they also live in difficult environments where there might be natural disasters, civil war and rapid population growth. Together this means that they have a low standard of living and a poor quality of life.

Poor countries find that it is impossible to catch up with the richer countries. Indeed, many of them find that they are falling further and further behind and getting increasingly poorer. This is because they are caught in the so-called **cycle of poverty** (shown in diagram B and on page 115). People living in poor countries often face a daily battle just to survive. To them, the future appears to have little hope. One such person is Marietta, who lives in south-east Kenya. Her daily lifestyle is described in diagram C.

A

GLOBAL NEWS

World poverty in 2013

- 1.4 billion people live in extreme poverty.
- They earn less than US$1 a day, which is about £200 a year.
- 38,000 children die each year from poverty.
- 23 of the world's 25 poorest countries are in Africa.

B

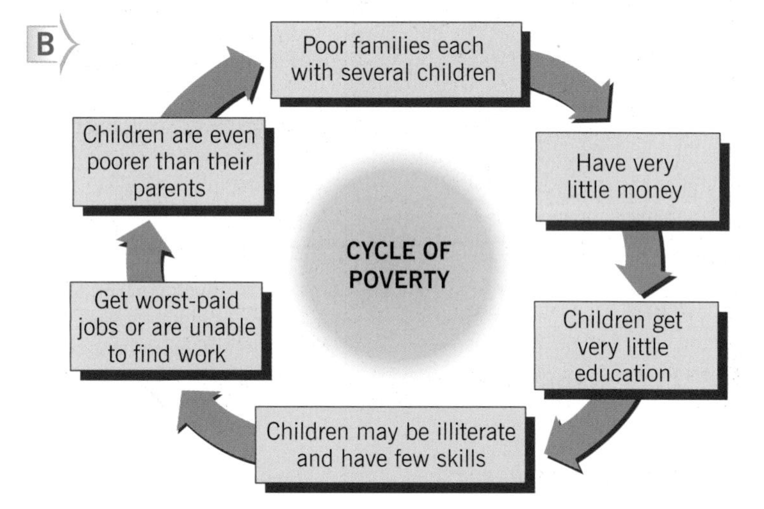

C

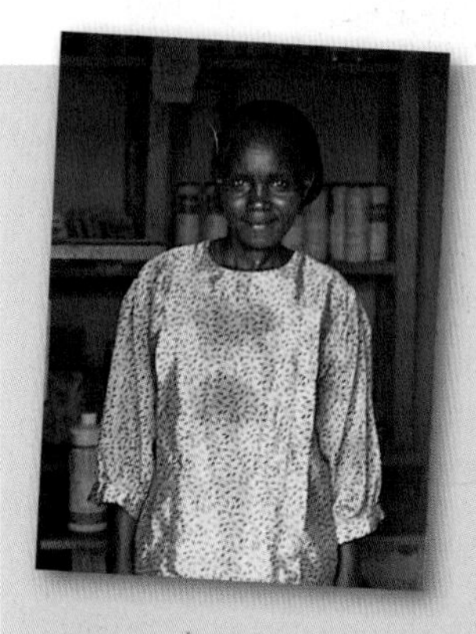

Marietta lives with her seven children, all aged under 12 years. Her husband works 250 km away in Mombasa but he never earns enough either to travel home or to send any money. With her nearest neighbour living 3 km away, Marietta is left alone to look after her small *shamba* (farm).

Her day begins as soon as it is light. She collects wood as this is the only source of energy for both cooking and warmth. The eldest girls have to go 2 km to the nearest river to collect the day's supply of water before school. They then walk another 5 km to school (there is no transport). Marietta spends most of her day collecting firewood and looking after her crops of maize, beans and sorghum. She also keeps 12 chickens, 20 goats and two cows. The cows are her main source of wealth. They provide milk and are used to plough the hard, dry ground. It is essential that they remain healthy as the nearest vet is over 50 km away and his bill would be too much for Marietta to pay.

Some causes and effects of poverty

D

Activities

1 a What is meant by 'the cycle of poverty'?
 b Make a larger copy of diagram **E**. Put the following into the correct boxes to show the effects of the cycle of poverty.
 - Family becomes even poorer
 - Family cannot afford to visit the very few doctors in the country
 - Family becomes weaker and are not well enough to work
 - Family likely to fall ill
 c Describe how Marietta is caught in the cycle of poverty.
 d How is daily life for Marietta and her daughters different from that of your family?

2 Look at the statistics on the back cover.
 a Name the two poorest countries.
 b With the help of diagrams **D** and **F**, list seven problems caused by poverty in these countries.
 c Describe any five of these problems.

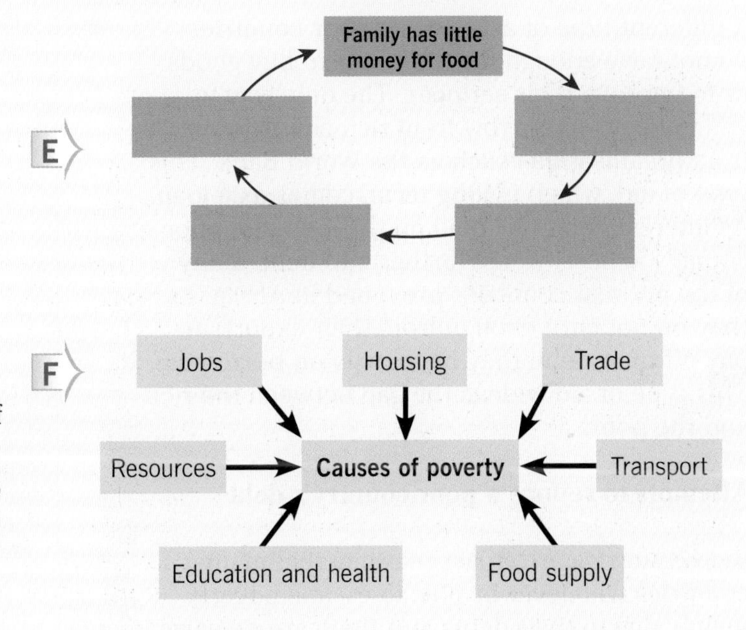

Summary

Poverty is a problem that affects large numbers of people around the world. It is difficult for people living in these countries to break out of the cycle of poverty.

How might poverty be reduced?

A major problem facing the world today is how to reduce poverty in poor countries and how to improve the standard of living and quality of life of people living there. This means trying to find ways to help people living there to break free from the cycle of poverty described on page 100. This can only be done with the help of the rich countries and worldwide organisations.

Countries remain poor when either:

- they fall into **debt** because the money they earn from their exports is less than the money they have to pay for their imports, or
- they borrow money in order to build things like schools, hospitals and roads.

Aid – a traditional method of help

Aid is the giving of resources such as money, medical equipment, food supplies or human helpers to another country. If you or your school are asked to give help it is likely to be for an emergency following a natural disaster such as an earthquake, flood or tsunami. This type of aid is **short term** (photo A).

A different type of aid is when poor countries do not have enough money either to buy goods or to improve their services. The only way to get money is to borrow from rich countries or from organisations such as the World Bank. This type of aid, which is **long term**, comes as a loan. As interest has to be paid on the loan, the poor country is likely to fall further into debt. Many of the poorest countries are found in Africa. As they do not earn enough from their exports to pay off their debt, they have to go on borrowing. This type of aid widens the gap between the rich and the poor.

Attempts to reduce a poor country's debt

Rich countries are at last realising that many countries will never be able to escape poverty mainly due to their debts, but there are no easy solutions. Five suggestions are given in drawing B.

A Evacuating patients after a natural disaster

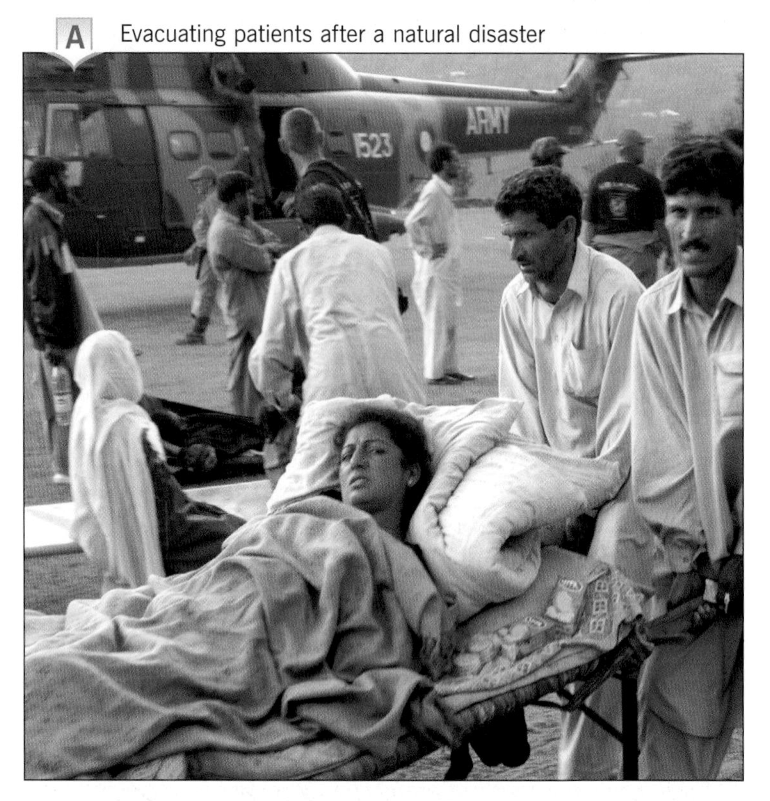

B How can a poor country's debt be reduced?

Self-help – an alternative approach

Practical Action, which used to be known as the Intermediate Technology Development Group, is a British charity. It has a different approach to overcoming poverty in poor countries. It believes that aid is of limited value as it is often only given for a short period of time. Practical Action suggest that it is better to encourage people to look after themselves in order to gain self-respect and independence. It believes in the proverb: 'Give a man a fish and you can feed him for one day [aid], teach him how to fish and he can feed himself for life [education]'. If he is also taught how to make fishing nets then he becomes independent.

Practical Action has helped people in countries like Kenya through projects such as those shown in diagram C. Marietta, who we read about on page 100, has also been helped by Practical Action (diagram D).

C

Marietta was chosen by her community to become a vet. She then received basic training by Practical Action. A five-day course helped her recognise and treat simple livestock diseases and enabled her to give basic animal health care. She was then given an animal first-aid kit.

Each week she spends two mornings in her local 'surgery' and other days visiting local *shambas*. To reach the surgery Marietta has a four-hour return trek along a track where lions are known to have killed people! Some *shambas* are even further away.

She earns a small commission from the sale of vaccines and medicine, but does not receive a salary. However, she can now keep her two cows healthy and has some money to buy food in times of shortage.

D

Activities

1 a What is meant by the term 'aid'?
  b What is the difference between short-term aid and long-term aid?
  c Give two advantages and two disadvantages of each type of aid.

2 a Make a copy of drawing E.
  b Add labels to show how a charity organisation like Practical Action can help reduce the effects of poverty.
  c Describe how Practical Action has helped Marietta. Use the following headings:
  - Training/education
  - Benefits to herself
  - Benefits to others

E

Summary

There are no easy solutions when trying to reduce poverty in the world's poorest countries. One way is for people to help themselves – this is called self-help.

Trade is important in the world because it helps countries share resources and earn money. Unfortunately, not all countries get a fair deal from world trade. For example, poorer countries tend to **export** mainly primary goods which earn little money but **import** mainly manufactured goods that cost a lot of money. This unfair trade is a main reason why poor countries remain poor and many of their people live in poverty.

In this enquiry you work for an organisation that supports **fair trade**. Fair trade can help people in poorer countries to make more money and escape from poverty. This can help improve their standard of living and quality of life. The main aims of the organisation are shown in drawing B.

Your organisation has been approached by Kenya Coffee, a company that has decided to support fair trade. Kenya Coffee buys coffee beans mainly from Kenya (photo A). It ships the beans to the UK where they are processed and sold throughout Europe. Your task is to explain the main points of fair trade to the company and suggest how it can become a fair trade company.

A

B

Works for a better deal for Third World producers

- Encourages fair trade between countries
- Aims to reduce poverty mainly by paying higher prices to producers
- Tries to ensure a safe and healthy working environment for workers
- Supports sustainable farming methods and encourages a concern for the environment

How can fair trade help reduce poverty?

1 Look at drawing D.
   a How much money goes to the coffee growers?
   b How much money ends up in Kenya?
   c How much money ends up in the UK?
   d If the price of coffee in the shops goes up, where does most of the extra money go to?

2 Look at drawing E. List the points which may help:
   a reduce poverty
   b improve conditions for growers
   c protect the environment
   d affect people in the UK.

   Not all of the points in the drawing need be used.

3 Write a report for Kenya Coffee about fair trade, using your answers to Activities 1 and 2. Use the headings shown in diagram C. Link any suggestions you make to the aims of fair trade in drawing B.

C

The fair trade report

- The need for fair trade
- What a fair trade company needs to do
- What the coffee growers will give in return
- The effects of fair trade on people living in the UK
- The effects of introducing fair trade

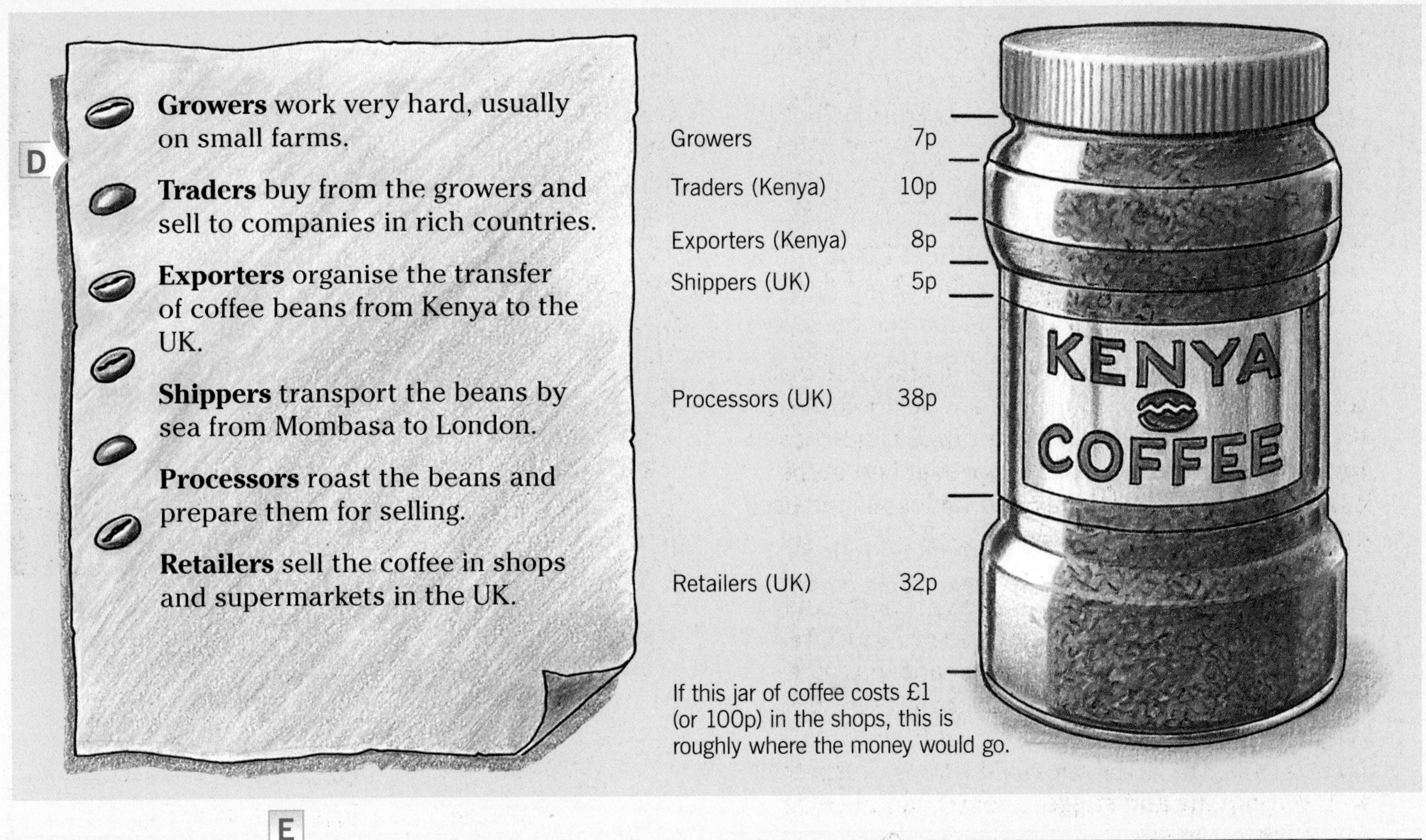
D
Growers work very hard, usually on small farms.
Traders buy from the growers and sell to companies in rich countries.
Exporters organise the transfer of coffee beans from Kenya to the UK.
Shippers transport the beans by sea from Mombasa to London.
Processors roast the beans and prepare them for selling.
Retailers sell the coffee in shops and supermarkets in the UK.
Growers 7p
Traders (Kenya) 10p
Exporters (Kenya) 8p
Shippers (UK) 5p
Processors (UK) 38p
Retailers (UK) 32p
KENYA COFFEE
If this jar of coffee costs £1 (or 100p) in the shops, this is roughly where the money would go.

E
Pay higher than the world market price for coffee beans.
Pay money to farmers in advance to prevent debt.
We will feel good about helping reduce poverty.
If we are giving the coffee growers all this, in return they must make some promises.
Insist on minimum health and safety standards.
Guarantee equal rights for women and no child labour.
Use sustainable farming methods.
Some jobs may be lost.
Increase the price of coffee in the shops.
Fair Trade
Works for a better deal for Third World producers
Set up processing factories in Kenya.
Ensure decent wages and good housing.
Coffee will cost a bit more.
Minimise damage to the environment.
Buy direct from the growers.

How can we use an atlas?

An **atlas** is a book containing many different maps. Some maps show very large areas, such as the entire world or whole continents. Others show smaller areas such as a country or part of a country. Atlases also contain a number of different types of map. These include **physical maps**, **political maps** and a variety of other maps showing many different features as in drawing A. Maps like these are called thematic maps because they show a particular topic, or theme, in a specific area.

Atlases have changed over recent years. Many now show much more data on maps and include diagrams and tables of statistics.

To use an atlas properly you need to look carefully at the **contents** page. This can be found at the front of the atlas and gives the page number for every map, table and diagram in the book. The **index** at the back of the book shows exactly where a particular place may be found.

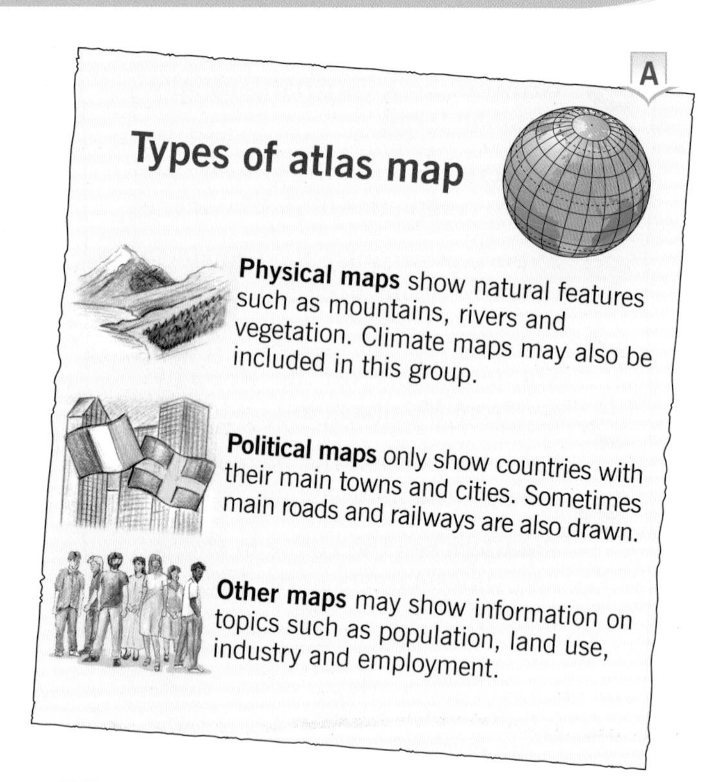

Activities

1. **a** Look at the contents page shown in drawing **D**. Sort the maps under the headings:
   - Physical
   - Political
   - Others

   b Use an atlas to add three further maps to each group.

2. Look at map **C**.

   a How many countries are there in South America?

   b For each country, name the capital city.

3. **a** Copy and complete table **E** below for Buenos Aires, Lima, La Paz, Manaus and São Paulo. Use the information on atlas maps **B**, **C**, **F** and **G**.

   b Use an atlas to collect the same information about three places in Europe, including where you live.

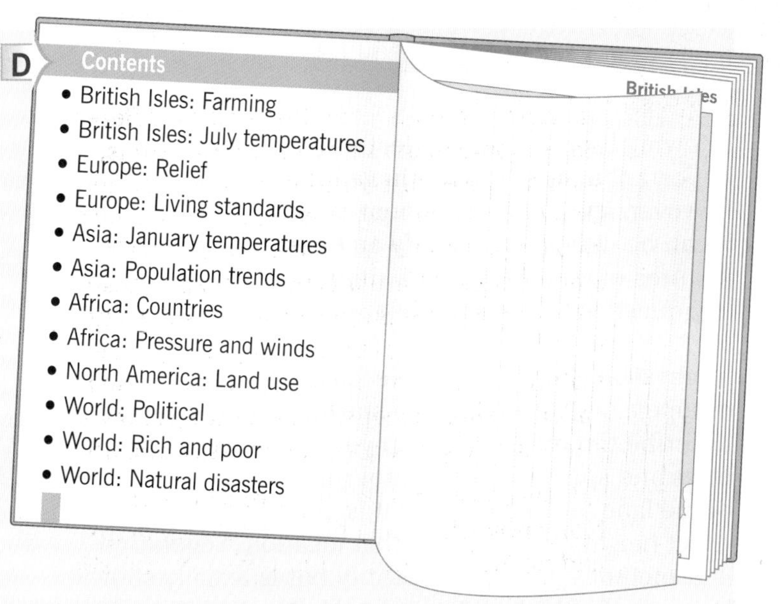

D

Contents

- British Isles: Farming
- British Isles: July temperatures
- Europe: Relief
- Europe: Living standards
- Asia: January temperatures
- Asia: Population trends
- Africa: Countries
- Africa: Pressure and winds
- North America: Land use
- World: Political
- World: Rich and poor
- World: Natural disasters

E

Town	Country	Natural vegetation	Annual rainfall
Buenos Aires	*Argentina*	*Grassland*	
Lima			

Summary

An atlas provides us with all sorts of information, both physical and human. Using the contents and index helps us make the best use of an atlas.

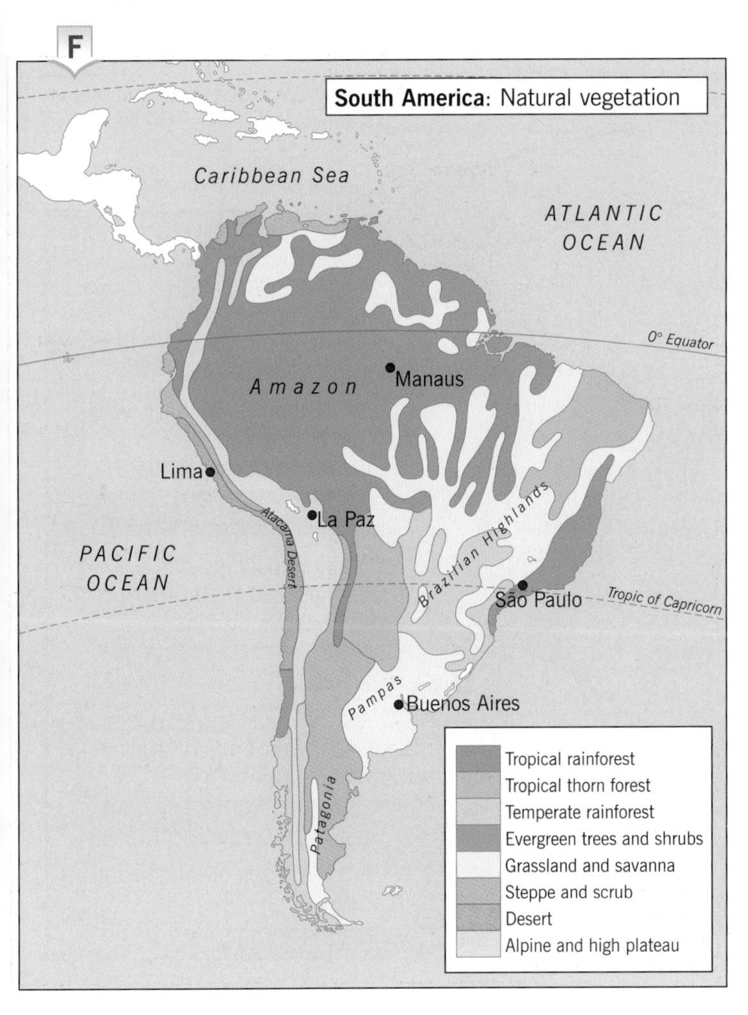

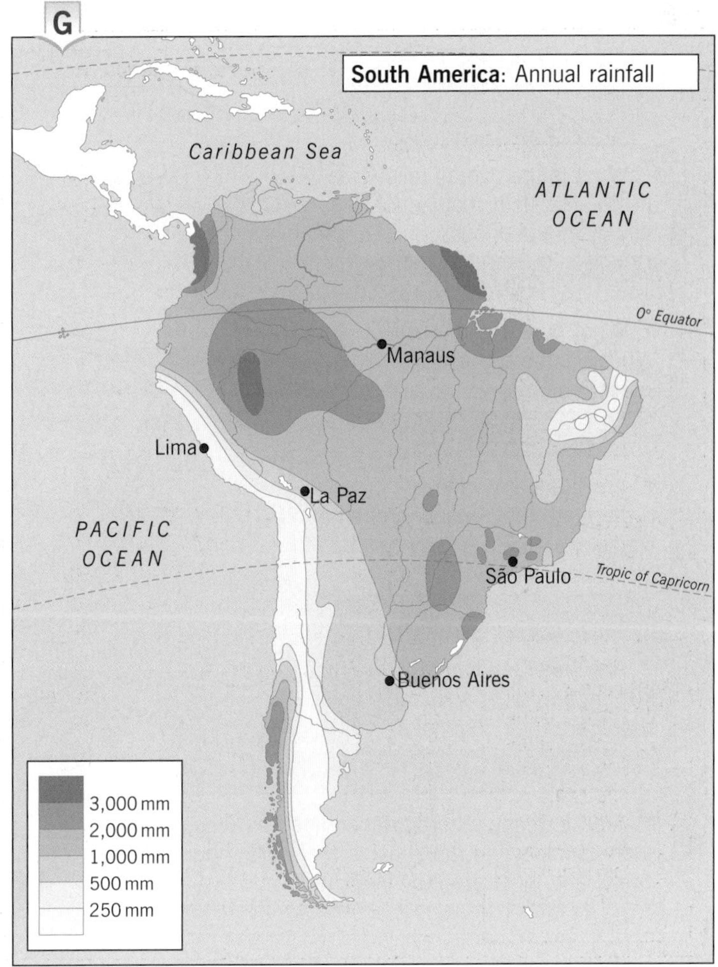

How can we describe physical features on a map?

Knowing how to describe features on a map is an important and useful skill in geography. Maps, however, show a large amount of information and it can be difficult to identify and describe the main features of an area. Using a simple checklist or set of key questions can make the task easier.

In **physical geography** we need to describe the relief, drainage and vegetation, as shown in drawing **A**. To describe the **relief** on a map you need to use **contour lines** and **spot heights**. Contour lines show the height of the land and what shape it is. Spot heights give the exact height of the land at that location. **Vegetation** is sometimes difficult to identify but is usually shown by symbols which are explained in the map key.

Look at the key questions below and see how you could use them to describe the physical features of map **C** opposite.

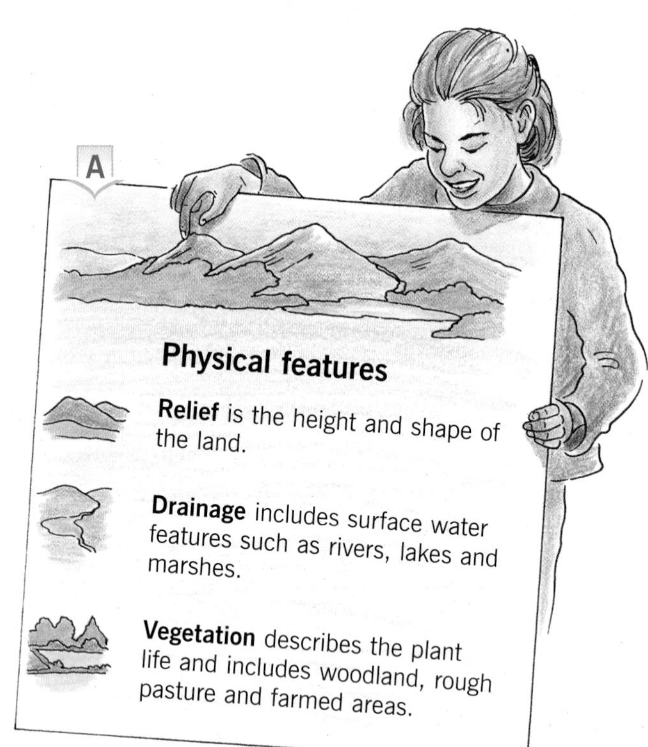

C 1:25,000 Ordnance Survey map showing part of the Lake District

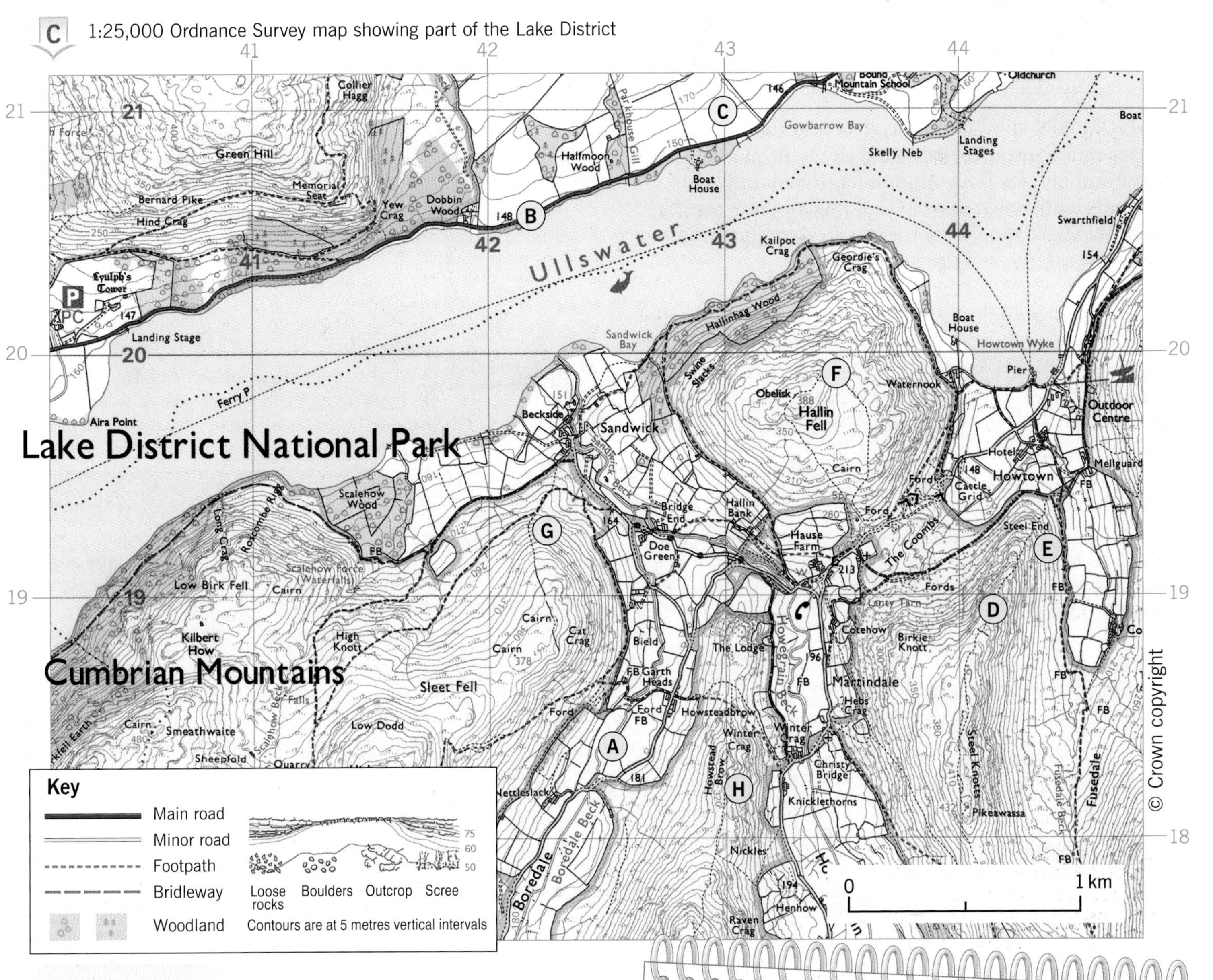

Activities

1. Look at the eight places circled on map **C**. Match the letters to each of the following relief features:
   1. Gentle slope
   2. Steep slope
   3. Narrow ridge
   4. Rocky outcrops
   5. Flat valley floor
   6. Steep-sided valley
   7. Round hill
   8. Smooth shoreline

2. Read description **D** of the area shown on map **C** above. Copy and complete the description using the following words:

straight, hills, smooth, uneven, streams, bays, 500, moorland, ridges, steeply, woodland

D

The area is part of the Lake District and is rugged and mountainous. The land is generally .... sloping and is a mixture of rounded .... and narrow .... The highest peak, High Dodd, is over .... metres in height.

Most of the valleys in the area are .... and steep-sided. The southern part of the area is drained by three .... which flow into Ullswater. Most of the area is rough .... with a few small areas of .... The north shoreline of Ullswater is mainly .... The south is more .... and has some ....

Summary

Physical features on a map include relief, drainage and vegetation. They may be described using a simple checklist.

How can we describe human features on a map?

Maps usually show both physical and human features. **Physical features** are the natural part of the environment, such as mountains, rivers and lakes. **Human features** are those that have been made by people. These include settlement, communications and how people use the land.

As we have seen on page 108, a simple checklist or set of key questions can make the task of describing map features much easier. Look at the key questions below and see how you could use them to describe the human features on maps **B** and **C**.

A

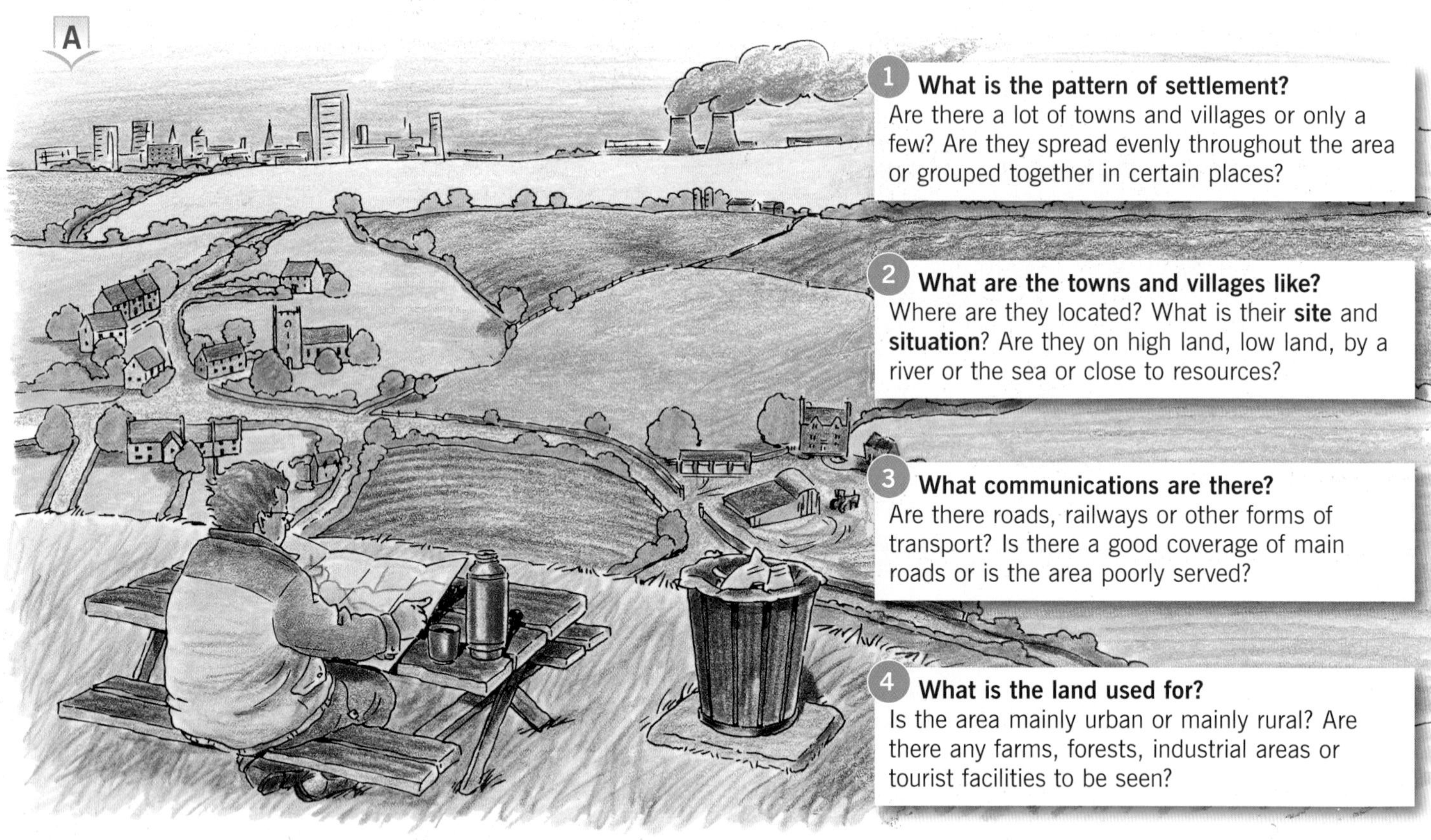

B

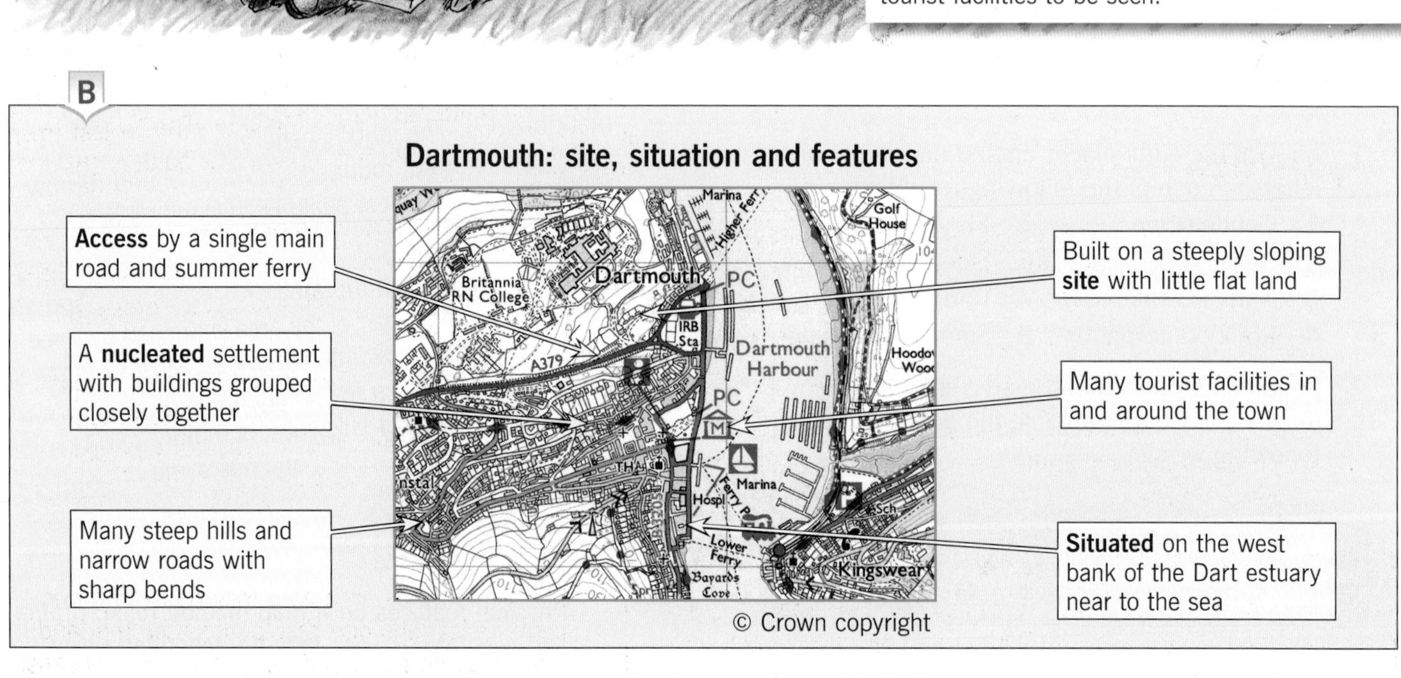

C 1:25,000 Ordnance Survey map showing part of South Devon

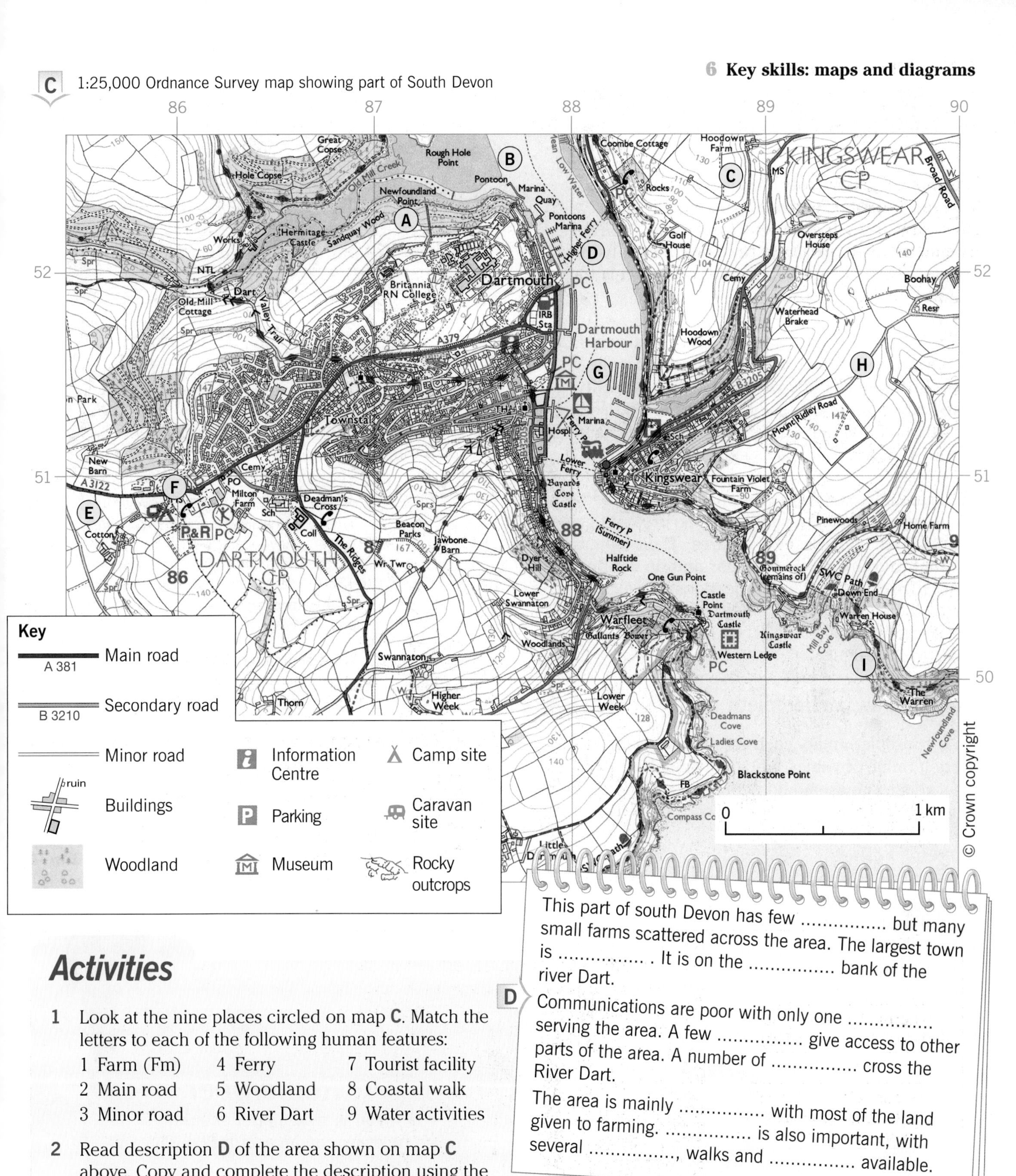

Activities

1 Look at the nine places circled on map **C**. Match the letters to each of the following human features:

1 Farm (Fm)	4 Ferry	7 Tourist facility
2 Main road	5 Woodland	8 Coastal walk
3 Minor road	6 River Dart	9 Water activities

2 Read description **D** of the area shown on map **C** above. Copy and complete the description using the following words:

D

This part of south Devon has few .... but many small farms scattered across the area. The largest town is .... It is on the .... bank of the river Dart.

Communications are poor with only one .... serving the area. A few .... give access to other parts of the area. A number of .... cross the River Dart.

The area is mainly .... with most of the land given to farming. .... is also important, with several ...., walks and .... available.

- minor roads
- car parks
- camp sites
- villages
- west
- main road
- ferries
- Dartmouth
- rural
- tourism

Summary

Human features on a map include settlement, communications and how people use the land. They may be described using a simple checklist.

What do choropleth maps show?

Choropleth maps use different colours or shading to show variations between places. Map A is a choropleth map. It uses colours to show differences in rainfall totals in Britain. Notice how clearly the pattern of rainfall is shown and how easy it is to compare rainfall amounts in different areas.

Choropleth maps, or **graded shading maps** as they are also called, should be used when you want to show **differences** between places and to identify **distribution patterns**. They are often used to show variations in population density or differences in wealth within a country, as shown on map B.

The main advantages of choropleth maps are that they are easy to draw, simple to interpret and give a good overall impression of an area. Their limitations are that they are based on areas and do not provide exact figures for particular locations. For example, on map A, London could have anything between 750 mm of rain and nothing at all. The map simply does not give an exact figure.

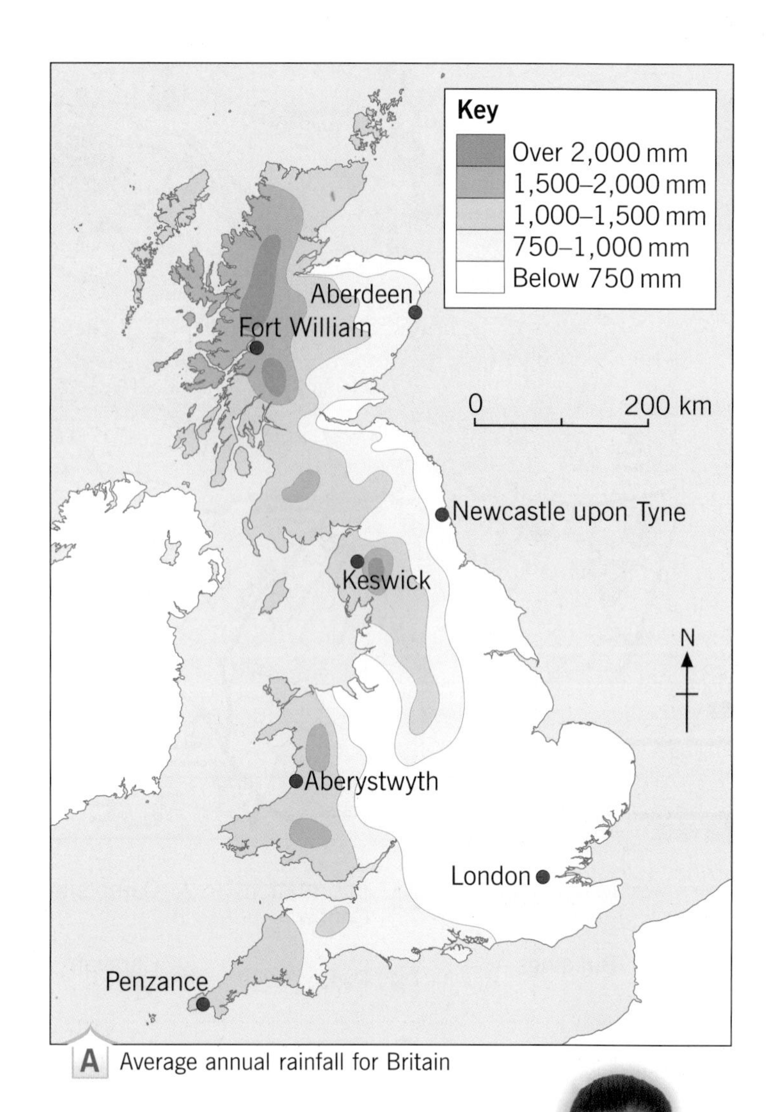

A Average annual rainfall for Britain

How to draw a choropleth map

On a choropleth map, each colour stands for a certain range of values. The first thing is to decide what those values should be. Look carefully at the figures and divide them into equal groups. Four to six is usually the best.

Next decide how to colour or shade the groups. Shading should be in just one or two colours and graded from dark to light. Areas with the highest values should be darkest and those with the lowest values should be lightest.

Finally draw the map. You can do this by following the instructions on map B.

✔ Remember

When you draw a map it should:

- ✔ be drawn on an accurate outline or traced
- ✔ be drawn in pencil so that any mistakes can be easily corrected
- ✔ have a key that gives the meaning of any symbols or colours used
- ✔ have a title and labels that are printed in pen.

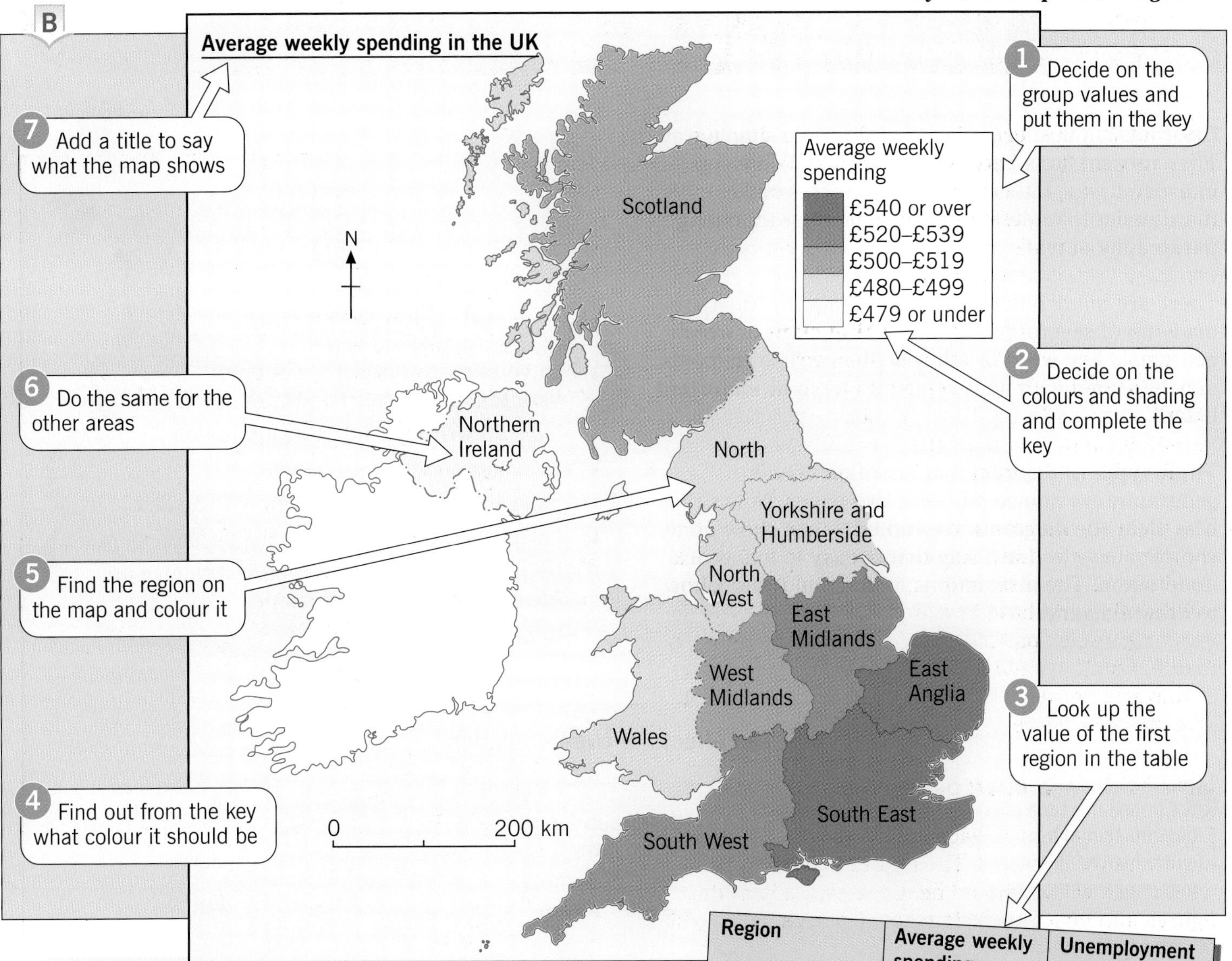

C

Region	Average weekly spending	Unemployment rate (%)
North	£480	10.1
Yorkshire and Humberside	£491	8.9
North West	£499	7.9
East Midlands	£530	7.8
West Midlands	£507	9.4
East Anglia	£583	6.7
South East	£614	6.6
South West	£529	6.2
Scotland	£518	7.1
Northern Ireland	£469	7.8
Wales	£480	8.4

Activities

1 Look at map **A**.
   a Name the towns with over 1,000 mm of rainfall.
   b Add *east* or *west* to these sentences:
   - Britain's wettest area is the north ....
   - Britain's driest area is the south ....
   - Rainfall decreases from ...... to ....

2 Describe the pattern of wealth in the UK.

3 Draw a choropleth map to show unemployment rates in the UK. Use an outline of map **B**, the data in table **C** and the key in drawing **D**.

D

- 10 and over
- 9–9.9
- 8–8.9
- 7–7.9
- 6.9 and under

Summary

Choropleth maps use graded shading to show information about places. They show distribution and may be used to identify patterns.

How can we use diagrams in geography?

Diagrams show information in a clear and simple way. They present facts and explain ideas and concepts in a visual way, rather like a picture. This makes them easier to understand and remember than long paragraphs of text.

There are many different types of diagram. Most are made up of several parts or elements, each of which contains a **key word** or a **key sentence**. The elements may be joined with lines or arrows to show important links.

Three types of diagram that are often used in geography are shown on these two pages. Notice how clear the diagrams are and how they show quite complicated ideas in a way that is easy to follow and understand. The instructions in diagram **A** show how to draw a diagram.

✔ Remember

When you draw a diagram, it should:

✔ have text that is short and carefully worded
✔ have all writing printed neatly in pen
✔ be coloured lightly in pencil
✔ have all straight lines drawn with a ruler
✔ have a title which says what the diagram shows
✔ be sketched on rough paper first to give an idea of layout, size and shape.

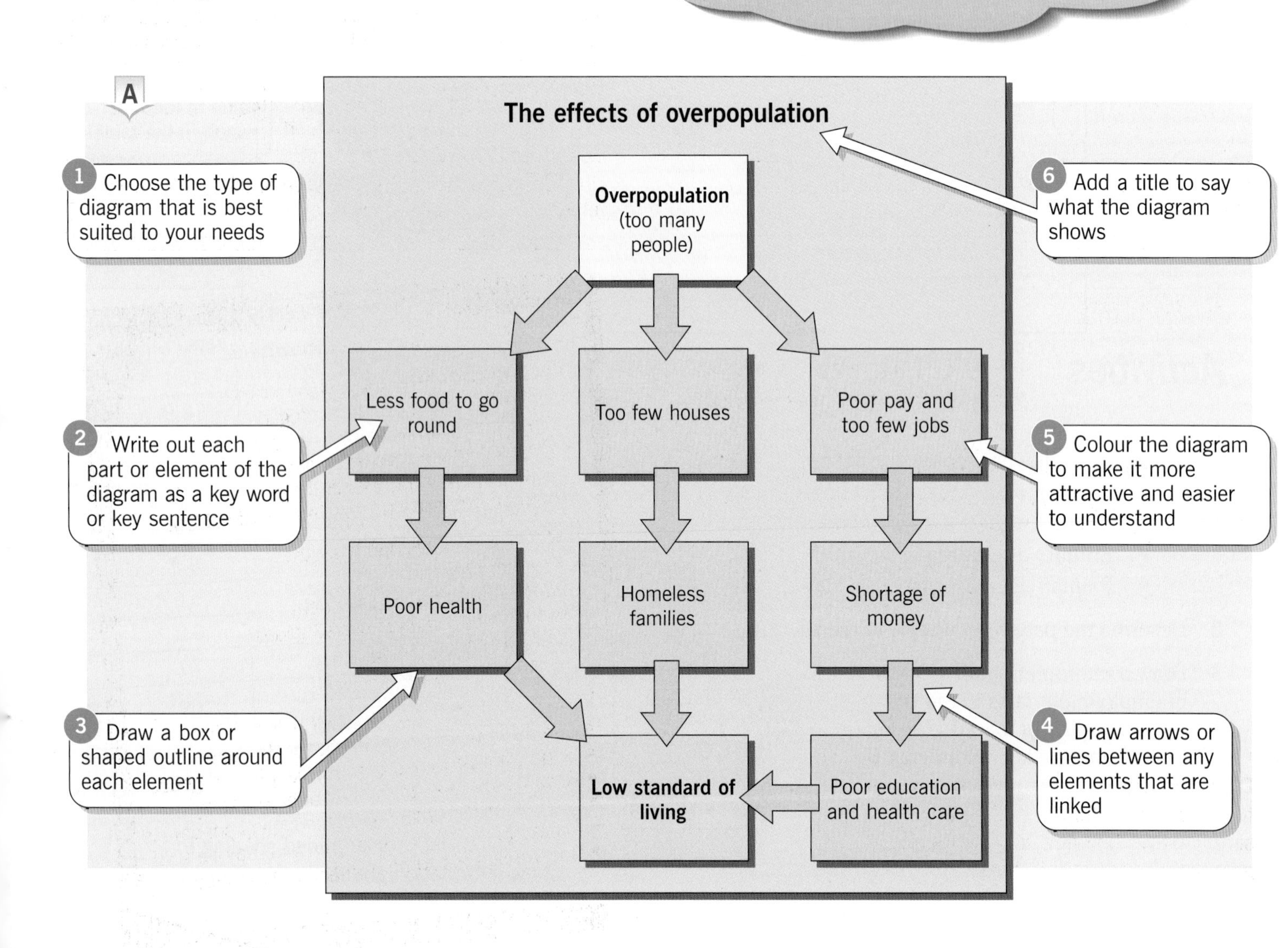

B

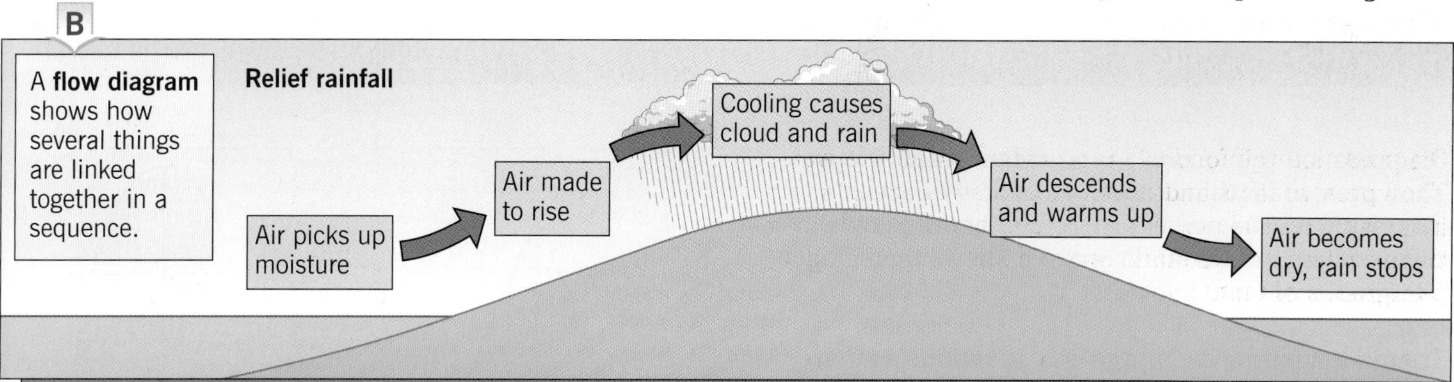

A **flow diagram** shows how several things are linked together in a sequence.

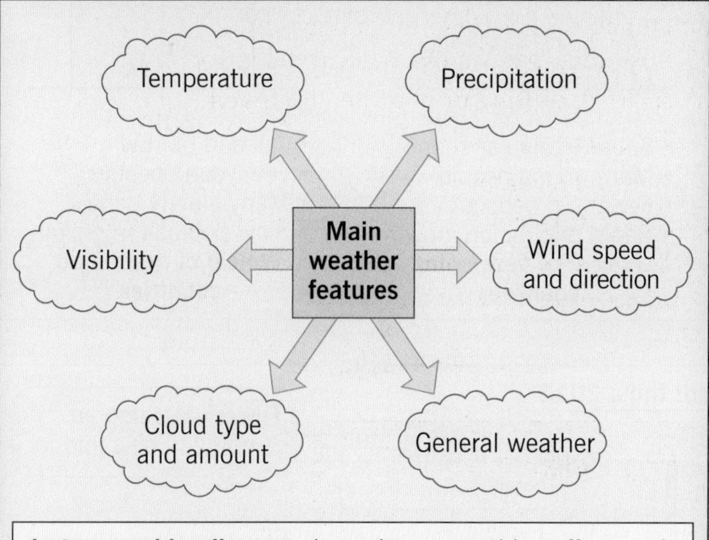

A **star** or **spider diagram** shows how something affects or is affected by several different factors.

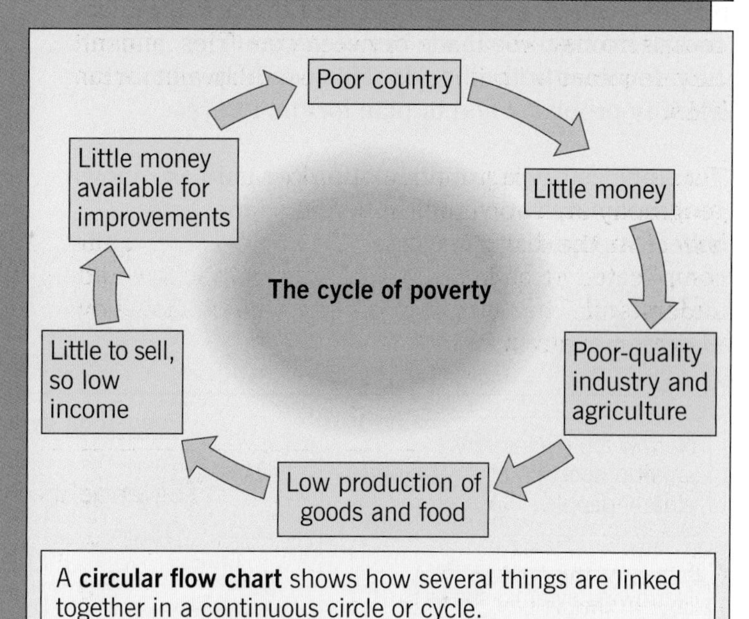

A **circular flow chart** shows how several things are linked together in a continuous circle or cycle.

Activities

1. Draw a star diagram to show the advantages of using diagrams in geography. Try to give at least six points.
2. Draw three diagrams to show the information in diagram **C**. Follow the instructions numbered ❶ to ❻ in diagram **A**. You may need to put the parts or elements in the correct order.
3. Draw a diagram to show the main stages of your journey to school. Start at home and finish in your classroom.

Summary

Diagrams use key words and key sentences to show information in a clear and simple way. They help us to understand geographical facts and ideas.

C

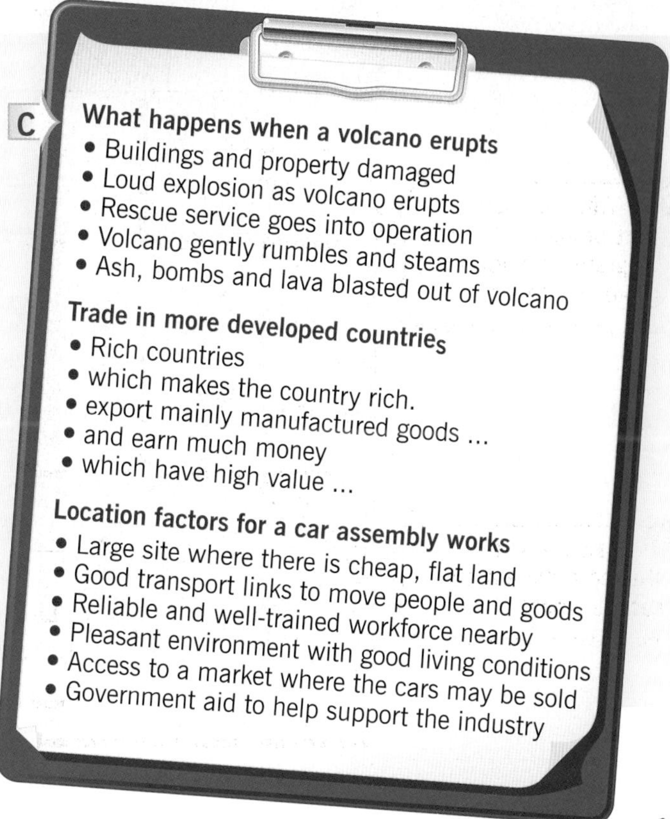

What are population pyramids?

A population pyramid is a type of **bar graph** that shows two main things about a country or area. First it shows the percentage or number of people in different age groups, and second it shows the balance between males and females.

Population pyramids, or **age–sex pyramids** as they are also called, are useful because they enable **comparisons** to be made between countries, and help **forecast** future trends. This can help a country identify problems and to plan for the future.

Two different types of population pyramid are shown in diagram **A**. Notice their different shapes and different population features.

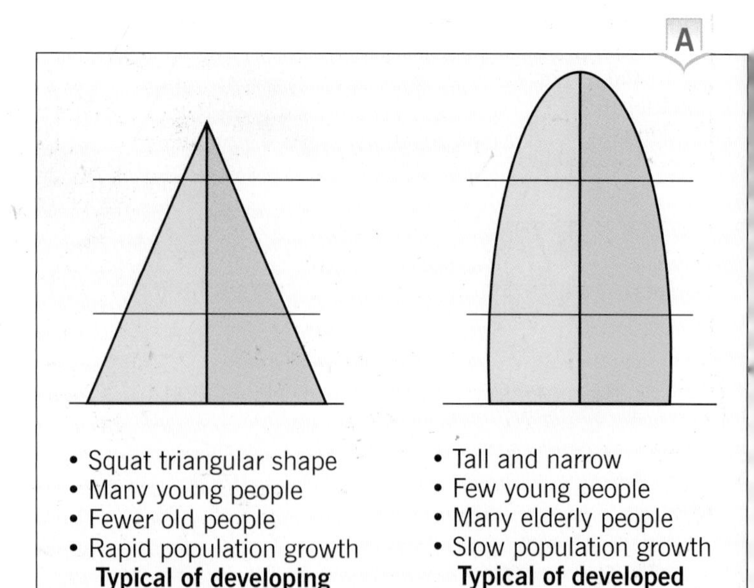

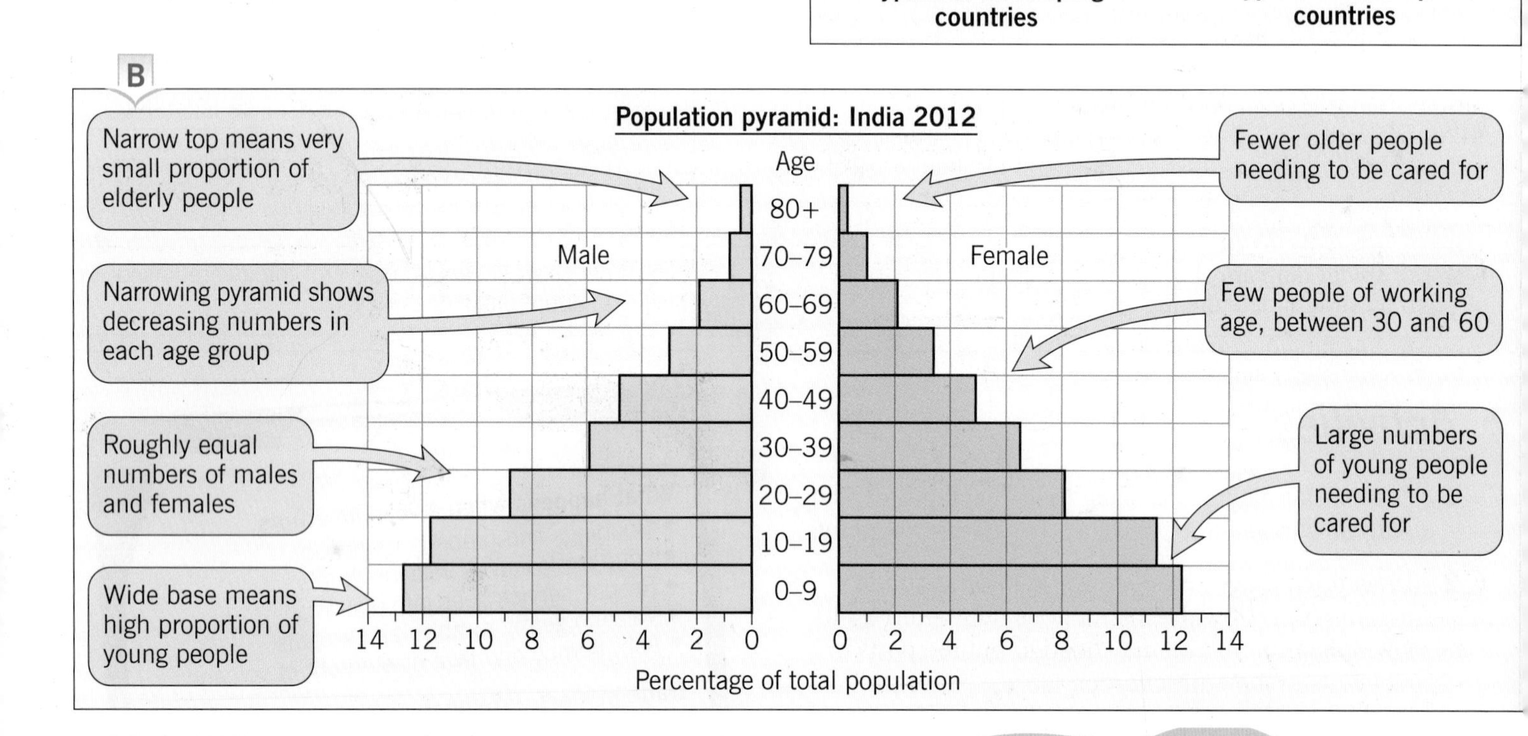

Activities

1 a What are the two main things that a population pyramid shows?

b Why are population pyramids useful?

2 List the statements from drawing **C** that best describe what graph **B** shows about India's population.

C

- The graph is tall and narrow.
- The graph is triangular in shape.
- There are decreasing numbers in each age group.
- There are almost equal numbers in each age group.
- 25% of the population is less than 10 years old.
- There is a low proportion of young people.
- There is a small proportion of older people.
- There is a high proportion of young people.
- There are few elderly people.
- Male and female numbers are similar.

How to draw a population pyramid

Population pyramids are simply two bar graphs that are drawn on either side of a vertical axis. They are best drawn on graph paper, or on a population pyramid graph outline if there is one available.

If you are drawing two or more graphs to compare countries or areas, they should be drawn at the same scale. The male side is usually drawn on the left and coloured blue. The female side is on the right and coloured red or pink.

When you draw a population pyramid it should:

- ✔ be between a half and a third the size of a page
- ✔ be drawn in pencil so that any mistakes can be corrected easily
- ✔ have all straight lines drawn with a ruler
- ✔ have a title and labels that are printed in pen.

D

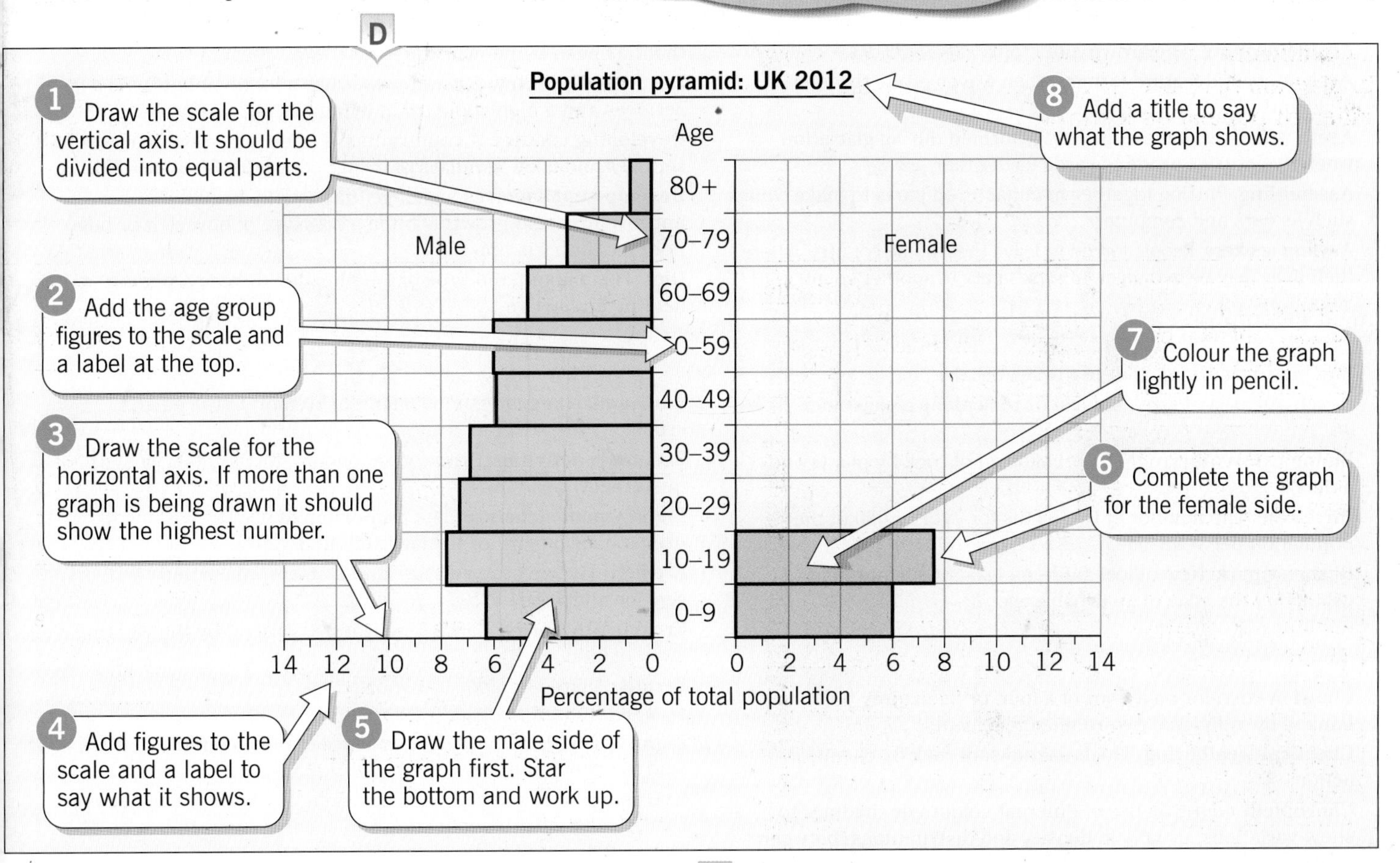

E Population data for the UK in 2012

Age group	% males	% females
80+	0.8	1.8
70–79	3.2	4.4
60–69	4.7	5.6
50–59	6.0	6.0
40–49	5.9	5.8
30–39	6.9	6.9
20–29	7.3	7.0
10–19	7.8	7.6
0–9	6.3	6.0

3 a Copy and complete graph **D** using data from table **E**.

b List the statements from drawing **C** that best describe what your completed graph shows about the UK's population.

c Is the UK's population growth likely to be slow or rapid? Give reasons for your answer.

Summary

A population pyramid compares the number or percentage of people in different age groups. It can help forecast future population growth.

Glossary and Index

Abrasion Erosion caused by the rubbing of material carried by rivers, glaciers, waves and or wind. *8, 22*

Age–sex pyramid The proportion of males and females in selected age groups and shown as a pyramid. *78, 116–17*

Agriculture The growing of crops and rearing of animals. *30–33, 83, 100*

Aid Help given by rich countries, international agencies and voluntary organisations to poorer countries. It may be short-term or long-term. *102*

Alluvium Sometimes called silt, it is fine material left behind after a river floods. *15*

Arable farming The growing of crops. *30–31, 33, 93*

Arch A coastal feature formed when waves erode through a headland. *16*

Arête A narrow, knife-edged ridge formed during glaciation when two corries erode towards each other. *23–24*

Assembling Putting together manufactured parts to make things such as cars and computers. *36–37*

Asylum seekers People forced to leave their home country, where their lives may be in danger, to seek safety in another country. *61*

Bay A wide, curving inlet of the sea or lake. *16, 46*

Beach An area of sand, shingle or rock along a coastline. *17, 20, 46*

Biological weathering The breakdown of rock by plants and animals. *7*

Birth rate The number of live births for every 1,000 of the population per year. *56–57, 78, 81*

Business park New offices built in pleasant surroundings, usually on the edge of an urban area. *42–43*

Chawl A corridor on a floor of a four- or five-storey building flanked by individual two-roomed tenements. *75, 81*

Chemical weathering The breakdown of rock by chemical action. *7*

Choropleth map The use of different colours or shadings to show variations, usually in density and distributions, between places. *112–13*

Cliff A steep, rocky slope that may overlook the sea. *16, 18–19*

Climate The average weather condition of a place taken over a number of years. *22, 30–33, 40, 53, 74, 76–77, 90–93, 96, 107*

Climate change How the average annual temperature or annual rainfall changes over a period of time. *90–93, 98*

Contour A line drawn on a map to join places of the same height above sea level. *46, 108*

Corrie A deep, rounded hollow with a steep back wall and sides formed by glacial erosion. *22–24*

Crofting A type of farming found in north-west Scotland which produces food on a small scale. *30*

Cross-section A view at right-angles across a landform. *14*

Current The flow of water in a certain direction. *8, 14, 17*

Cycle of poverty When the level of poverty increases year after year with little hope of improvement. *100, 102, 115*

Death rate The number of deaths for every 1,000 of the population per year. *56–57*

Debt Where poorer developing countries owe money to richer developed countries or international organisations. *102*

Densely populated An area that is crowded with people. *50–54, 79*

Deposition The laying down of material carried by rivers, sea, ice or wind. *9, 14–15, 22, 24–25*

Deposition landforms Landscape features made up of material that has been laid down. *9, 14–15, 17–18, 24–25*

Developed country A country that has a lot of money, many services and a high standard of living. *38, 84, 86, 98, 102, 116*

Developing country A country that is often quite poor, has few services and a low standard of living. *38, 40, 55, 74–87, 96–105, 116*

Development Involves a change that usually brings improvement and growth – often a measure of how rich or how poor a country is. *86–87*

Diet The amount and type of food needed to keep a person healthy and active. *87, 98–99, 101*

Earthquake A sudden movement, or tremor, of the earth's crust. *71, 76*

Economic activities Primary, secondary, tertiary (service) or quaternary jobs. *26–47*

Employment structure The proportion of people working in primary, secondary or tertiary activities. *29*

Erosion The wearing away of land by material carried by rivers, sea, ice and wind. *6, 8–14, 16–20, 22-25*

Erosion landforms Landscape features formed by the wearing away of land. *8, 10–14, 16, 22–25*

Erratic A large boulder transported by ice and deposited in an area of different rock. *25*

Exports Goods and services produced in one country that are sold to other countries. *84–85, 101–102, 104*

Extreme poverty When people, lacking the basic living needs of food, health, shelter, education and security, have to live on $1 or less a day. *100–101*

Factory A place where things are made from natural resources and raw materials. *34–37*

Fair trade When poor countries are allowed to compete with rich countries to help them escape the cycle of poverty. *104–105*

Flood plain The flat land at the bottom of a river valley which is often flooded. *15, 77, 79*

Fold mountains Land formed when two plates move together, pushing up the land between them. *76*

Fossil fuels Non-renewable energy resources such as coal, oil and natural gas that come from the fossilised remains of plants and animals. *90, 94–95*

Freeze–thaw A form of physical weathering in which rock is split by water in cracks repeatedly freezing and thawing. *6, 19, 22*

Glacial trough A deep valley with steep sides, a flat floor and a U-shaped cross-section formed by a glacier. *24*

Glacier A slow-moving tongue of ice that flows down former river valleys under gravity. *8, 22–25, 92*

Global warming The warming of the earth's atmosphere, believed to result from the burning of fossil fuels which release carbon dioxide. *90–93, 96*

Gorge A steep-sided valley. *12*

Greenfield site Land found on the edge of an urban area which has not yet been built on. *42-3*

Greenhouse effect The way that gases in the atmosphere trap some of the heat from the sun. *90–91*

Gross national product (GNP) The wealth of a country and a measure of its economic development. It is the total amount of money earned by a country in a year divided by its total population. *86–87, back cover*

Hanging valley A tributary valley left high above the main valley as its glacier was unable to erode downwards as quickly as the larger glacier in the main valley. *24–25*

Headland A part of the coastline that juts out into the sea and usually ends in a cliff. *16*

Health How fit and active, or vulnerable to disease, a person is. *81, 83, 86–87, 93, 97–99, 101*

High-tech industries Industries such as computing and telecommunications that use advanced techniques and skill to make high-quality goods. *29, 42–45*

Human features These have been made by people and include settlements, transport and the use made of the land. *30, 46–47, 110*

Ice age A very cold period of climate change when glaciers covered large areas of land and ice created distinctive landforms. *19, 90*

Immigrant A person who arrives in a country with the intention of living there. *60–63*

Imports Goods bought by a country that were produced in other countries. *35, 84–85, 102, 104*

Industrialised A country with highly developed industries that uses advanced machinery and skilled workers. *35*

Industry Any type of economic activity that produces goods or provides a service. *28, 34–37, 42–45, 53*

Interdependence When countries work together and rely on others for help. *84*

International migration The movement of people from one country to another. *59*

Labour Workers, employed people. *34–37*

Landscape Scenery – what a place looks like. *6*

Life expectancy The average age a person born at a given time can expect to live. *78, 87, back cover*

Literacy rate The proportion of people that can read and write. *73, 87, back cover*

Load Material carried by a river. *11*

Long-term aid A type of aid given by richer countries to help poorer ones to develop and improve their standard of living. *102*

Malnutrition Ill-health caused by a shortage of food or a poor diet. *98–99, back cover*

Manufacturing Industries that make or assemble goods such as steel and cars. *28, 34–37, 42, 84, 104*

Market A group of people who buy raw materials and goods; or a place where the raw materials and goods are sold. *34–37*

Meander A large bend in a river. *5, 14–15*

Migrant A person who moves from one place to another to live or work. *58–62*

Migration The movement of people from one place to another to live or work. *58–63*

Mixed farming The growing of crops and the rearing of animals in the same area. *30–31*

Monsoon Meaning a season, each year there is a wet and a dry one in India. *74, 76–77, 82*

Moraine Loose rock that is transported, and later deposited, by a glacier. *22, 24*

Multicultural society When people of different ethnic groups, religions, languages and customs live and work together. *62*

Natural resources Raw materials that are obtained from the environment, e.g. coal, trees and water. *28, 35, 73, 94*

Negative factors Things that discourage people from living or working in a place. *52–53, 66–67*

Obese A person who is overweight. *98–99, back cover*

Onion-skin weathering The breakdown of rock by alternate heating and cooling which causes the surface layers to peel off. *6*

Ordnance Survey The official government organisation responsible for producing maps in the UK. *37, 46, 109, 111*

Pastoral farming The rearing of animals. *30–32, 100*

Physical features The result of natural processes and events that create landforms e.g. rivers, glaciers and volcanic eruptions. *46–47, 108, 110*

Physical map Shows natural features such as rivers and mountains as well as climate and vegetation. *106*

Plate tectonics The concept that the earth's crust is divided into a number segments that move around in relation to each. *76*

Plucking A process of glacial erosion when ice freezes onto rock and pulls some of it away as the glacier moves. *22*

Plunge pool A hollow at the base of a waterfall caused by erosion. *12*

Political maps Show human-created features such as countries, cities and railways. *106*

Pollution Noise, dirt and harmful substances produced by people and machines that spoil water, air and land. *73, 81, 91, 96–97*

Glossary and Index

Population density The number of people living in a given area. *50–51, 64, 79*

Population distribution How people are spread out over a given area. *50, 52, 54, 79*

Population explosion A sudden and rapid rise in the number of people. *56*

Population growth rate A measure of how quickly the number of people in an area increases. *56–57, 78*

Population pyramid The proportion of males and females in selected age groups and shown as a pyramid. *78, 116–17*

Positive factors Things that encourage people to live or work in an area. *52–53, 66–67*

Power The energy needed to work machines and to produce electricity. *34–35, 37, 94–95*

Primary industries Activities that collect and use natural resources, e.g. farming, fishing, forestry and mining. *28–33, 84, 104*

Pull factors Things that attract people to an area. *59*

Push factors Things that make people want to leave an area. *59*

Pyramidal peak A triangular-shaped mountain formed during glaciation by three or more corries eroding backwards towards each other. *23–24*

Quaternary industries High-tech activities that provide information and advice or are involved in research. *29, 42–45*

Raw materials Natural resources that are used to make things. *34–36, 53*

Relief The shape of the land surface and its height above sea-level. *30–33, 53, 71, 74, 106, 108*

Relief rainfall Rain caused by air being forced to rise over hills and mountains. *77, 115*

Ribbon lake A long, narrow lake found on the floor of a glaciated valley. *24–25*

Rural-to-urban migration The movement of people from the countryside to towns and cites. *59, 79, 96*

Science park An estate of modern offices and high-tech industries having links with a university. *42–43*

Sea defences Features added to a coast to protect it from erosion and flooding. *20–21, 93*

Secondary industries Activities that make, or manufacture, things by processing raw materials or assembling parts to make a finished product, e.g. steelmaking and car assembly. *28–29, 34–37, 84*

Service industries Occupations, also known as tertiary industries, that aim to help people, e.g. nursing, teaching, transport and retailing. *28–29*

Sewage Waste material from homes and industry. *81, 83, 96, 101*

Shanty settlement A collection of huts and poor-quality housing which often lack electricity, a water supply and means of sewage disposal. *80–81*

Short-term aid Emergency help given after a natural disaster such as an earthquake, tsunami or flood. *102*

Site The place where a settlement or factory is located. *34–37, 80, 110*

Sparsely populated An area that has few people living in it. *50–54, 79*

Spit An accumulation of sand or shingle that grows outwards from a coastline or across a river mouth. *17–18*

Spot height A point on a map giving its exact height above sea level. *46, 108*

Stack A pillar of rock on the coast detached from the mainland by erosion. *16*

Standard of living How well-off a person or a country is. *39, 72, 86, 100, 102, 104*

Sunrise industries Activities that are growing in importance, e.g. computers. *44*

Sunset industries Activities that are declining in importance, e.g. shipbuilding. *44*

Tertiary industries Activities that provide a service for people, e.g. nursing, teaching, transport and retailing. *28–29, 38–41*

Tourism When people travel to places for recreation and leisure. *38–41, 46–47*

Trade The movement and sale of goods between countries. *80, 84–85, 104*

Trade balance The difference between the cost of imports and the value of exports. *84*

Trade deficit When a country spends more money on its imports than it earns from its exports. *84*

Trade surplus When a country earns more money from its exports than it spends on its imports. *84*

Transport Ways of moving people and goods from one place to another. *34–38, 40, 43–45, 82, 101, 110*

Transportation The movement of material by rivers, sea, ice and wind. *9–10, 17–18, 22, 24–25*

Truncated spur A ridge of high land that used to extend into a valley but has had its end removed by a glacier. *24–25*

U-shaped valley A valley that has been eroded by a glacier so that its cross-section looks like the letter U. *24*

V-shaped valley A valley that has been eroded by a river so that its cross-section looks like the letter V. *10–11, 24*

Vegetation Describes the plant life of a place. *70–72, 92–93, 107–108,*

Waterfall Where a river flows over a sudden vertical drop. *4, 12–13, 24*

Water supply The availability of a continuous and reliable supply of clean water. *53, 81, 83, 96–97, 100–101, back cover*

Waves Formed when wind blows over the sea. *8, 16–19*

Weathering The breakdown of rocks by climate, chemicals, plants and animals. *6–8, 10, 19, 22*